gears and

transmissions

automotive technology series
VOLUME 4

GEARS AND TRANSMISSIONS

J. G. Giles, Ph.D., B.Sc.(Eng.), C.Eng., F.I.Mech.E., M.I.E.E.
with a chapter by D. Thaw, C.Eng., F.I.Mech.E., M.S.A.E., M.I.Mar.E.

LONDON
ILIFFE BOOKS LTD

Iliffe Books Ltd
42 Russell Square
London, W.C.1

First published in 1969

Filmset by Photoprint Plates Ltd.,
Wickford, Essex
Printed in England by
J. W. Arrowsmith Ltd., Bristol

592 00628 X

contents

general preface to automotive technology series

All the basic principles of Automobile Engineering are contained within normal engineering or scientific courses, but vehicle applications call for a mixture of scientific knowledge which goes to make up the specialist study. This text book series has been produced to collect the relevant theoretical and practical considerations in a number of volumes devoted to Vehicle Structures, Engines, Transmissions, Steering, Suspension and Tyres, Braking, Accessories, and Test Methods. Each volume contains contributions from highly qualified experts working in the respective fields and is written for engineers who are not experienced in that study. Space allocation is such that a good overall discussion of the relevant theoretical considerations combined with currect practice and possible future trends can be presented. References to literature for further study are given with each chapter, so that in addition to providing basic general knowledge of Automobile Engineering it can provide the entry to a specialist topic for an interested student.

Worcester, 1967 J.G.G.

preface to volume 4

Gears and Transmissions

The transmission of engine power to the driving wheels of a vehicle has always posed a number of problems to automobile engineers. Ideally, the full engine power should be available at all road speeds, but it has so far proved difficult to find a suitable mechanism to convert the engine power in this way and there are further problems introduced by the vehicle layout and the need to allow for movements of the suspension or steering.

The first chapter reviews some of these problems and the requirements of a vehicle transmission. Individual aspects of these requirements are covered in more detail in further chapters, together with the devices or mechanisms employed so that clutches and starting are considered together. Similarly gearboxes and gear-changing, automatic gearboxes and their control are covered in successive chapters. The use of overdrives and their effect on vehicle operating economy is considered, and this leads to the developing trend towards variable ratio transmissions which are described and discussed in a separate chapter, including the use of hydrostatic drives and the various 'shunt' arrangements needed for better efficiency. Finally, Mr. D. Thaw discusses the requirements for vehicle axles, which normally include a final reduction gear ratio needed to increase the gearbox or transmission torque to the values necessary at the driving wheels. All types of axles are illustrated including those with the 'limited slip' type of differential gear.

Worcester, 1968 J.G.G.

1

transmission layout and requirements

1.1 INTRODUCTION

A vehicle transmission is required to transfer the power developed by an engine to the driving wheels so as to propel the vehicle. Strictly speaking, the term 'transmission' includes all those parts situated between this engine and the wheels and employed for transmitting the engine power. In practice, however, many components are common to different vehicles and it is possible to speak of clutches, gearboxes, couplings, propeller shafts, axles and differentials so as to be more specific. There is a tendency, too, to refer to transmissions when considering a special type of gearbox or fluid drive.

The majority of the design requirements of transmission components arise from this basic duty of transferring engine power to do useful work at the road wheels. However, it must be remembered that the vehicle drives the engine when the vehicle is slowing down, descending gradients, etc. In other instances, brakes may be incorporated into the transmission line and extra duties are thus imposed on the gears, shafts and other components. Rough gear changes, engine vibration, and road shocks are other factors which must be considered in the design of transmissions.

There is much interest at the present time in transmission components as a result of different vehicle layouts which are becoming available. Advocates can be found for front wheel drive, rear-engine layouts and even four wheel drives for passenger cars. Similar trends are also noticed for passenger transport and commercial vehicles. These differing layouts have interesting repercussions on transmission components, some of which may disappear altogether whilst some quite new devices are in use. In particular, various types of constant velocity flexible joints are now available whereas much of the gearbox and driving axle may be combined in common casings, which may become part of the engine itself. The changes in transmission layout tend to introduce various space problems and limitations, and the difficulties are emphasised by the general trend towards automatic or variable ratio transmission which usually bring some further space or layout problems.

1.2 LAYOUT

Only recently has there been any move away from a vehicle layout which consisted of an engine mounted at the front of the vehicle driving through a gearbox and propeller shaft to a differential reduction gear installed in a rear axle casing. This particular layout was convenient in the earliest days of motoring especially for clutch and gearbox controls which were easily brought to the driver.

As a result of the general trend towards more compact vehicles with lower ground clearance etc., the intrusion of the gearbox and propeller shaft into the passenger compartment becomes more of a disadvantage than otherwise. The space requirement of a heavy rear axle and its movement during road travel are also difficult to satisfy. Hence there is a strong trend to do away with the vehicle's propeller shaft.

Clearly, if the driving axle and reduction gears can be brought closer to the engine, then a lot of weight, hence cost and space, can be saved. Two possible solutions which quickly come to mind are the front wheel drive, and the rear mounted engine versions which use the rear wheels to drive the vehicle. In both cases, it is possible to combine all the engine and power transmitting components into a single unit situated close to the driving wheels. Design, construction and service problems are affected in many ways by these new concepts of a vehicle, but once overcome there is much to be said for combining all the engine-transmission components in one place. It becomes possible to concentrate all effort on the rest of the vehicle to house its human, or other, load with the greatest convenience.

Enthusiasts for the front wheel drive version will remark on the French Citroen vehicles which have featured this layout for over thirty years and it has become a well proven and accepted design. As seen in Fig. 1.1 this vehicle layout uses the same engine, gearbox and axle components as for the more conventional arrangement, but they are connected together into single assembly which is then turned around to obtain a front wheel drive. The drive to the front wheels is carried by short shafts and flexible joints or universal couplings so that the springing and suspension of the wheels are not impaired.

More recently, the well-known BMC 'Mini' cars (Plate 1.1) have adopted an interesting front wheel drive layout in which the engine is turned across the vehicle so as to reduce the length of the engine-transmission compartment. In this arrangement, the gearbox is mounted alongside the crankshaft and, effectively, inside the engine crankcase. Because of its position, the axle reduction drive consists of ordinary helical gearing as compared to the usual bevel gears. The result is a very compact assembly containing the power transmitting components, and the success of this layout has caused

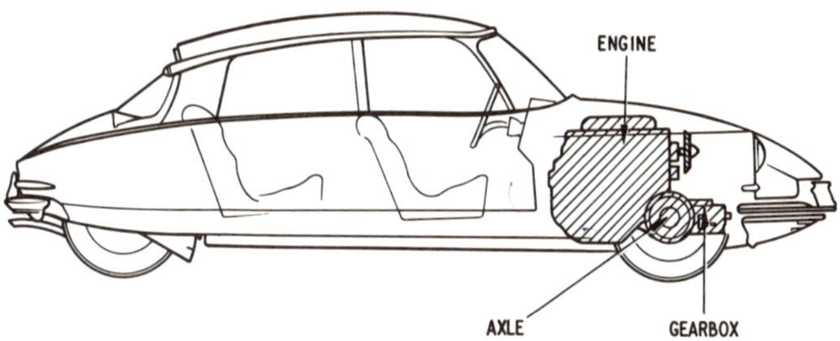

Fig. 1.1 The Citroen front wheel drive concept

much attention to be given to various alternative forms of 'power pack' or 'power plant'. The Triumph 1300 is another recent example of re-arranging components in an ingenious manner (see Fig. 1.2).

Future vehicle design, therefore, may be very much more inclined to the power pack concept. The power pack being an assembly occupying the minimum space to convert the vehicle's fuel into mechanical power in the form of rotating shafts suitable for connection to driving wheels.

Such power packs can be installed at the rear of a vehicle as well as at the front and there has been, perhaps, a greater tendency towards this rear-engine layout in vehicles manufactured on the Continent of Europe. The Volkswagen is one of the best-known examples of the rear-engine layout, the engine and gearbox being mounted in line but separated by the axle gearing. The drive into, and out of, the gearbox is made at the same end and required some revision of gearbox design. The mounting of the axle between engine and gearbox saves some space by utilising the flywheel and clutch housing for axle components. This arrangement is basically the same as for the Citroen front-drive layout.

In the rear-engined layout, the engine is mounted behind the axle for accessibility and this naturally tends to bring the centre of gravity of the vehicle towards the back of the vehicle. Similar layouts are employed on smaller Renault and Fiat vehicles and the

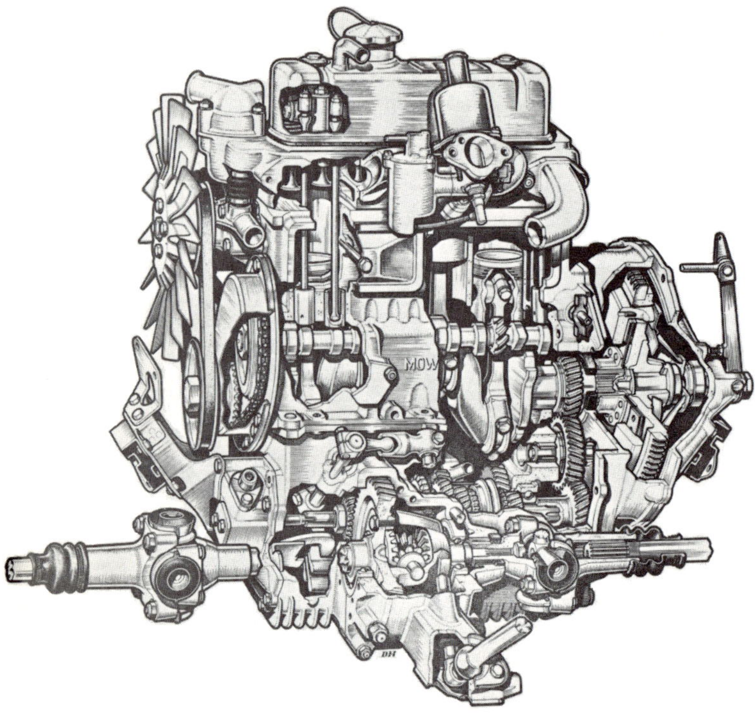

Plate 1.1 The BMC Mini transverse engine, gearbox and front wheel drive axle. (By courtesy of B.M.C. Ltd.)

Hillman Imp is a variant in which the gearbox input and output drives are not concentric. The Imp transmission layout is shown diagrammatically in Fig. 1.3 where it is seen that some simplification of the gearbox is obtained by positioning the input shaft to the gearbox above the axle pinion line.

The handling characteristics of the front-wheel drive and rear-engined vehicles are both different from those for the more conventional layouts, but either version seems to interest different

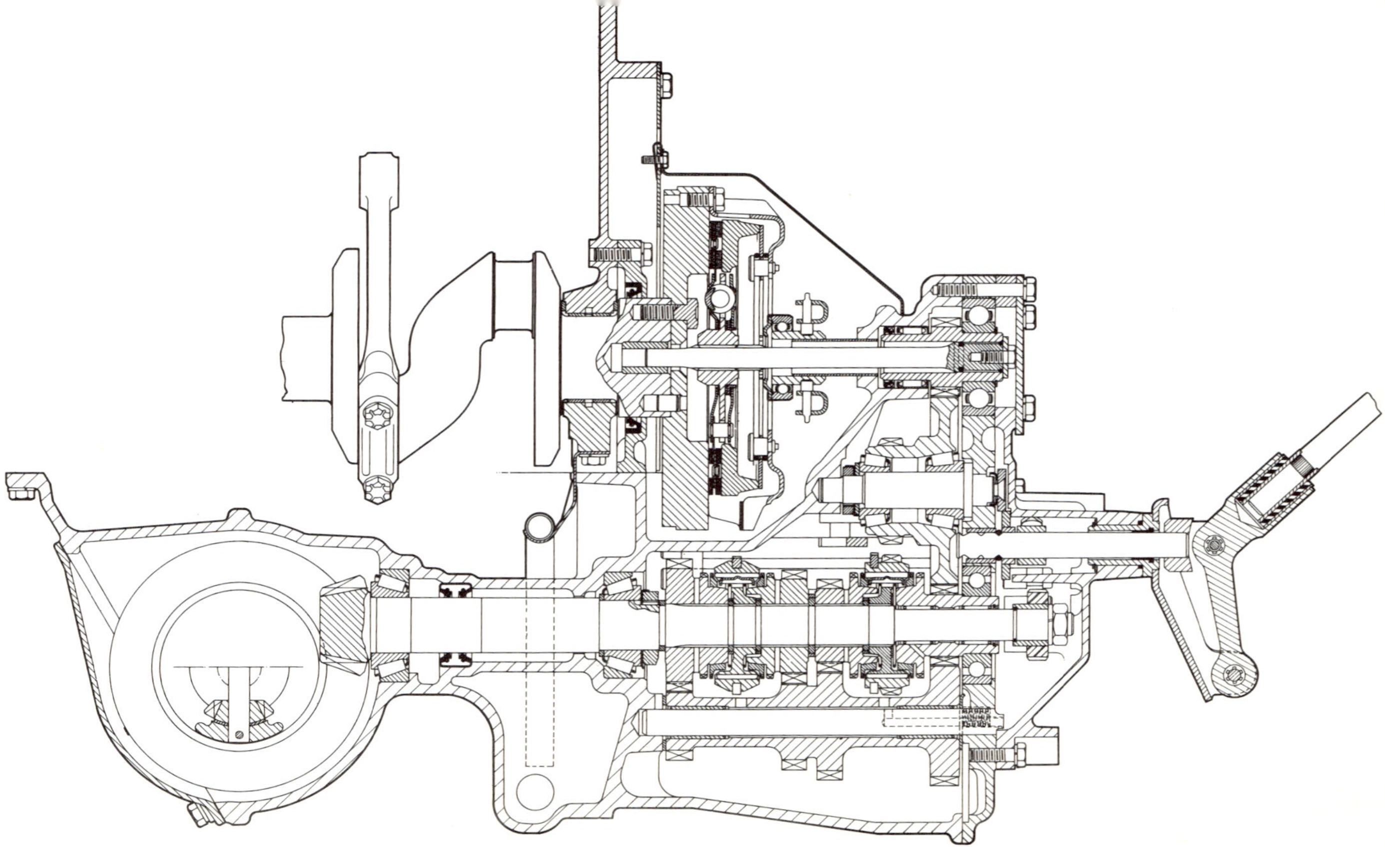

Fig. 1.2 A sectional view of the Triumph 1300 gearbox and axle. This longitudinal section of the transmission of the Triumph front wheel drive 1300 model shows that a single iron casing houses the primary drive gears, the four-speed, all synchromesh gearbox and the hypoid level fluid drive gears. The casting also forms the oil sump for the engine, and the complete assembly adds only a few inches to the height and length of the engine

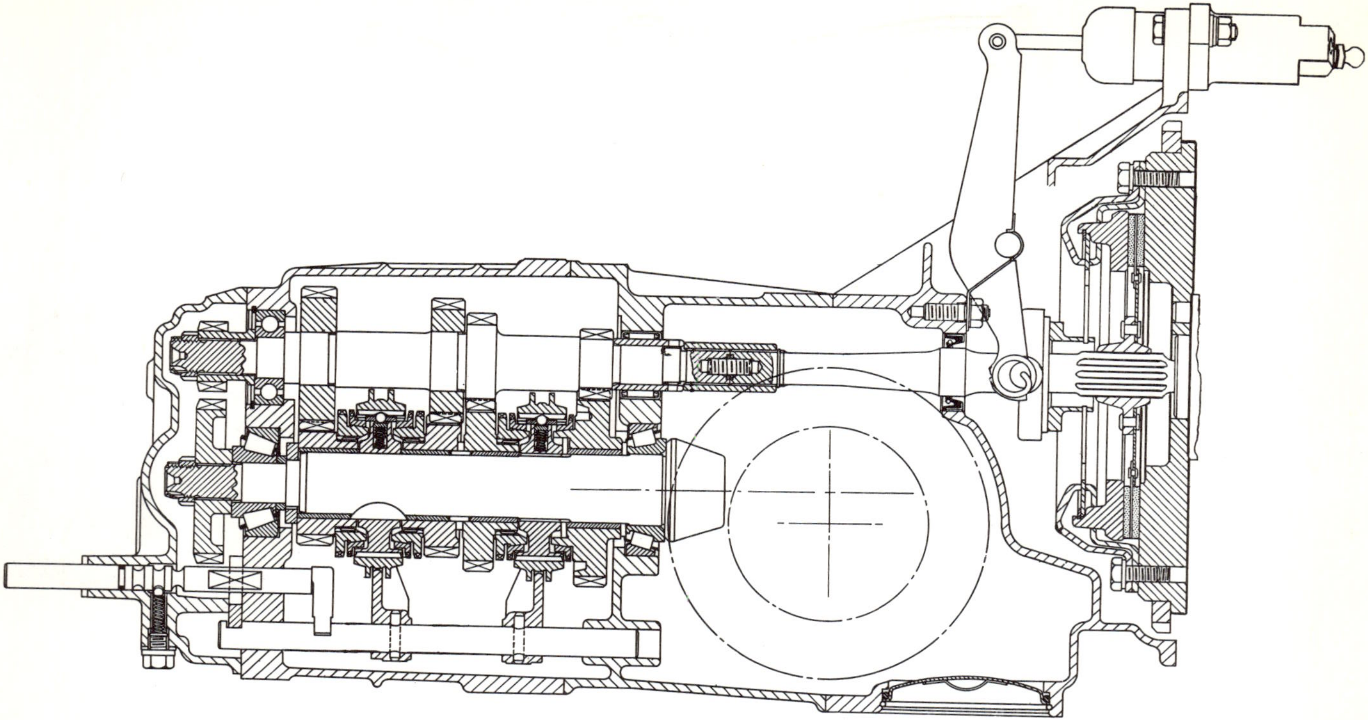

Fig. 1.3 A sectional view of the Hillman Imp transmission for a rear engine layout

drivers who praise the various characteristics or advantages. Four-wheel drive is available on a number of cross-country vehicles and is popular for various duties. The Ferguson transmission system (Fig. 1.4) included a four-wheel drive intended for normal passenger cars and in addition to its use in racing cars, has become available in the Jensen motor car.

An alternative approach which attempted to preserve the conventional layout was seen in some Pontiac compact cars. In these, the propeller shaft was replaced by a thin flexible shaft bent so that it was situated mainly beneath the floor line and thus avoided some of the space problems.

These examples are sufficient to show that the layout of transmissions and drives is very much affected by external requirements of vehicle concept, the type and nature of its load etc. Transmission designers must be ready to revise and re-arrange their major components to accept these requirements. The BMC Wolseley vehicle presented some difficulties of layout and one of the interesting features of the automatic transmission on the 1800 cm^3 engine was the introduction of a chain drive between the engine and transmission (see Plate 1.2). Ingenuity in producing satisfactory arrangements for new vehicle designs can reap great gains in the finished product.

The same arguments can be applied to commercial vehicles although changes in this class of vehicle occur less frequently. It may be said that the move towards the Power Pack is already in evidence from the use of tractor and trailers in heavy articulated vehicles. In passenger transport vehicles too, there are examples of under floor engines, and more recently, a number of rear-engined versions. Plates 1.3 and 1.4 show the layout of the engine-transmission for rear-engined buses which, by two alternative arrangements, achieve a compact assembly which can be removed as a single unit for maintenance, overhaul, etc. The separation of the vehicle structure from the drive unit obviously has many advantages for assembly and service. Moreover, it tends to separate two rather different types of engineering and design.

1.3 VEHICLE RESISTANCES

Having been presented with a proposed transmission layout for some new vehicle, a designer must then decide on the size of, and space needed for, his drive line. Power to drive the vehicle is transmitted mainly by mechanical parts whose sizes are dependent on torque or power which they must exert. The strength of gears,

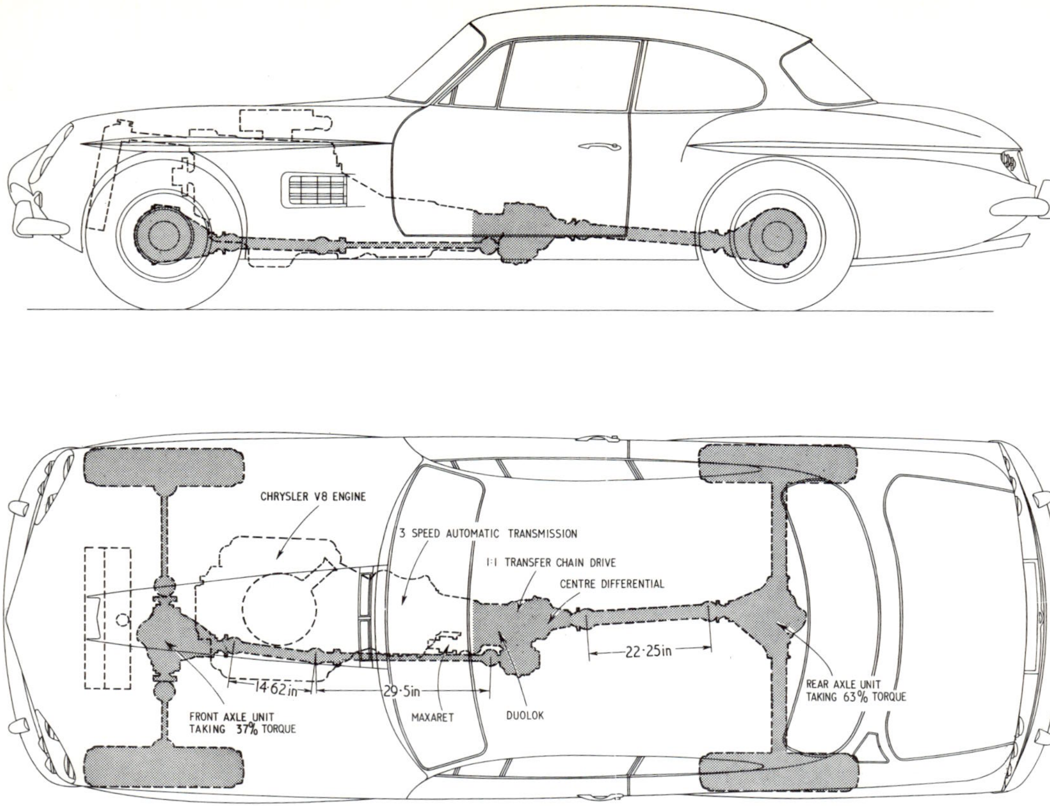
CHRYSLER V8 ENGINE
3 SPEED AUTOMATIC TRANSMISSION
1:1 TRANSFER CHAIN DRIVE
CENTRE DIFFERENTIAL
22·25in
REAR AXLE UNIT
TAKING 63% TORQUE
14·62in
29·5in
MAXARET
DUOLOK
FRONT AXLE UNIT
TAKING 37% TORQUE

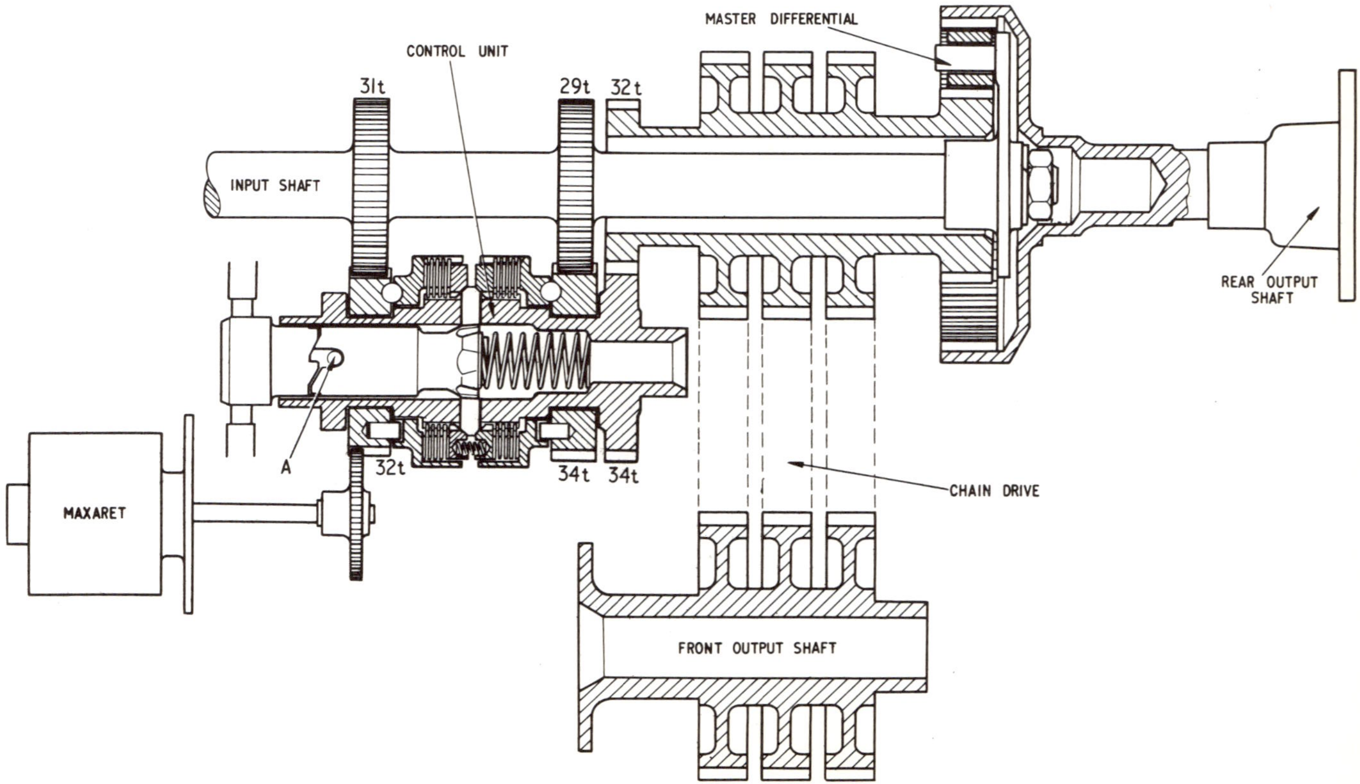

Fig. 1.4 The Ferguson gearbox and four-wheel drive unit

shafts and couplings must be sufficient to deal with the maximum engine conditions together with various overloads and inertia effects which may arise in the transmission line. Fortunately friction clutches, fluid devices, rubber mountings and tyres will slip at some overloads, and usually have a limiting effect on high

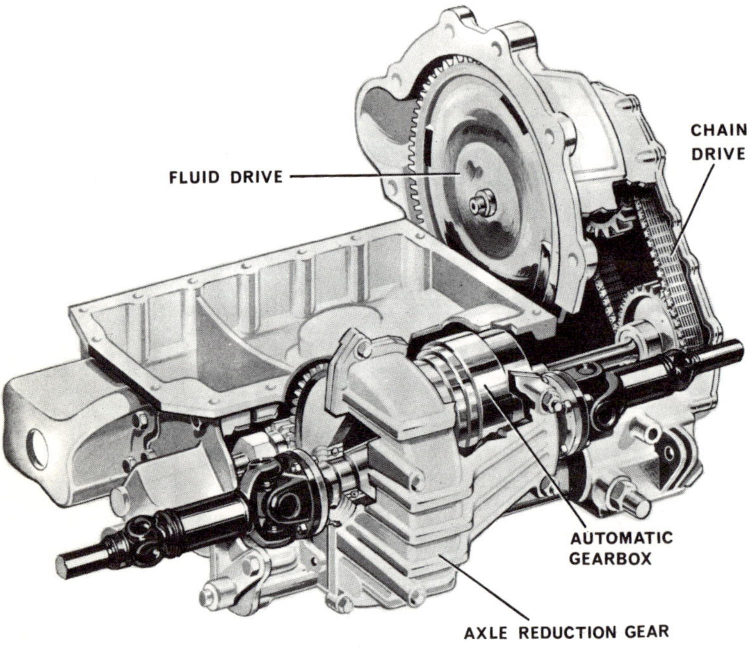

Plate 1.2 A chain drive is employed to connect a fluid drive on the engine to its automatic gearbox in the Wolseley 18/85 transverse engine layout. (By courtesy of B.M.C. Ltd.)

peak torques which can be developed, but it is usual to design mechanical components with some 'safety factor' in mind.

In practice, the vehicle does not operate for long spells at maximum power and spends the majority of its time at much lower power. The engine's power is absorbed in overcoming three basic resistances:

(1) Air resistance—due to the vehicle's motion through air (aerodynamic drag).

(2) Rolling resistance—due to tyres and bearings and depending on the vehicle load.

Plate 1.3 Rear engine layout for commercial vehicles. (By courtesy of Leyland Motors Ltd.)

(3) Gradients—work done against gravity by the vehicle when climbing a hill, or energy available when making descents.

1.3.1 AIR RESISTANCE

In displacing the earth's atmosphere, the vehicle expends some of the engine's power when moving along a road. The amount of drag, or resistance, increases rapidly with the speed of the vehicle and depends on the size and shape of the vehicle. For a given vehicle the aerodynamic drag will increase as the square of the vehicle speed and it is possible to express this drag in the following way:

$$\text{air resistance } R_a \text{ (lb)} = AKV^2 \qquad (1.1)$$

where A = cross sectional area of vehicle $(\text{ft})^2$, K = Shape Coefficient and V = Vehicle Speed (m.p.h.).

Thus it becomes possible to calculate the air resistance of a vehicle once its cross-sectional area and shape coefficient are known. Reasonable performance estimates can be made by assuming that at 50 m.p.h. the air resistance of a passenger car is equal to rolling resistance, but in commercial vehicles the air resistance is only about one half of the rolling resistance (R_r) at that speed.

1.3.2 ROLLING RESISTANCE

The resistance of the tyres and bearings of a vehicle in motion are approximately constant for a given load over a wide range of speeds. These resistances are rather higher when the vehicle is starting, and tyre resistances increase at speeds over 50–60 m.p.h. In general, however, a value of 40–50 lb per ton is a satisfactory figure to use in calculating vehicle rolling resistance. Many factors of tyre design, inflation pressures, road surface texture and surface material are responsible for this rolling resistance. Similarly, bearing drag depends on the load it is carrying, type of lubricant, operating temperature and other factors. As a result, it is difficult to predict this resistance accurately for a wide range of conditions and most engineers will leave a margin for error in any calculations.

1.3.3 GRADIENTS

In ascending a hill, the engine does work in hauling the vehicle up the gradient. Conversely, when the vehicle descends a gradient,

Plate 1.4 Rear engine layout for commercial vehicles. (By courtesy of Leyland Motors Ltd.)

the engine has to do less work or may be driven by the vehicle. Thus on a 1 in 10 gradient, a vehicle needs a 'push' of one tenth of its weight to enable it to ascend the hill in addition to the air and rolling resistances. On the down gradient, the vehicle can exert a 'pull' of one tenth of its weight against the other resistances. For convenience, in making estimates of performance and transmission calculations, engineers refer to hills in terms of the percentage of the vehicle's weight, i.e. 5, 8, 25% grades.

1.3.4 TOTAL RESISTANCE

The total resistance of a vehicle is the sum of these three resistances and is frequently known as 'road load'. In some cases a formula is used to predict this road load and the following is an example:

$$R_t = R_a + R_r + R_g \tag{1.2}$$

$$R_t = \frac{W}{100}(G+2) + AKV^2 \tag{1.3}$$

where R_t = total resistance (lb), R_a = air resistance (lb), R_r = rolling resistance (lb), R_g = gradient resistance, W = vehicle weight (lb) and G = percentage grade.

In Equation 1.3, the rolling resistance and gradient resistance have been combined into the form $W/100\ (G+2)$. The '2' term represents a rolling resistance of 1/50 of vehicle weight (44·8 lb/per ton of vehicle weight).

It is much more convenient as a rule to illustrate and predict performance by means of a graph of total resistance. Such a graph (Fig. 1.5) can show the road load over a range of speeds and, by constructing a series of curves, for various gradients. Fig. 1.5 shows the total resistances of a typical popular passenger car having a total weight of 1 ton. The rolling resistances are shown as a constant value of 45 lb for all road speeds, whereas the air resistance increases from zero when the vehicle is stopped to about 200 lb at 80 m.p.h. These combined resistances represent the total 'road load' for level road-calm air conditions. The dotted curves represent the total road load resistances for various up-hill gradients. Downhill gradients are not shown, but resistance curves could be constructed by subtracting the downhill 'pull' from the level road curve.

A vehicle's engine drives through the transmission to the road wheels and the above graph shows the thrust or tractive effort which must be developed at the wheels to maintain the vehicle at any given speed and hill gradient. Assuming an engine can develop 50 hp at the road wheels after allowance for various losses in the transmission, then this engine power would be represented by the chain-dotted curve which shows the thrust equivalent to 50 hp at each road speed. By noting the intersections with the various road

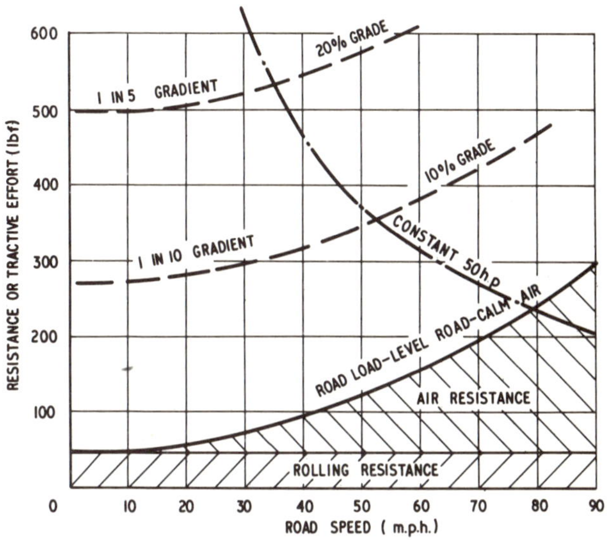

Fig. 1.5 Vehicle resistances and tractive effort curves

load curves, for instance, it is possible to deduce that this 50 hp will give the car a maximum speed of 78 m.p.h. on a level road. On a 10% grade the maximum speed would be reduced to 52 m.p.h. and at a 20% grade, the maximum engine power will be reached at 35 m.p.h.

A study of this graph shows that for most of normal driving at typical road speeds there is a very adequate reserve of tractive effort (and engine power) available. This reserve is, of course, employed for accelerating the vehicle when an increase in road speed is desired. The transmission designer's task is clear from this figure. He must arrange to convert the engine's power into tractive effort at the wheels over the full range of road speeds. A similar requirement exists for all types of vehicles which operate over a range of speeds. For commercial vehicles and tractors the road load curves will show a larger rolling resistance and a relatively smaller proportion due to air resistance, but the same type of diagram can be constructed.

1.4 POWER REQUIREMENTS

By converting the various resistances to power requirements at various speeds, it is easier to visualise the need for a range of gearbox or reduction ratios. An approximate formula derived from that for resistances is:

$$373\ P_t = \frac{WV}{100}(G+2)+AKV^3$$

where P_t = total power requirements (hp) (1.4)

Again a graph (Fig. 1.6) helps to visualise the overall requirement. The rolling resistances, having a roughly constant drag force, absorb more energy as the vehicle speed increases and are shown

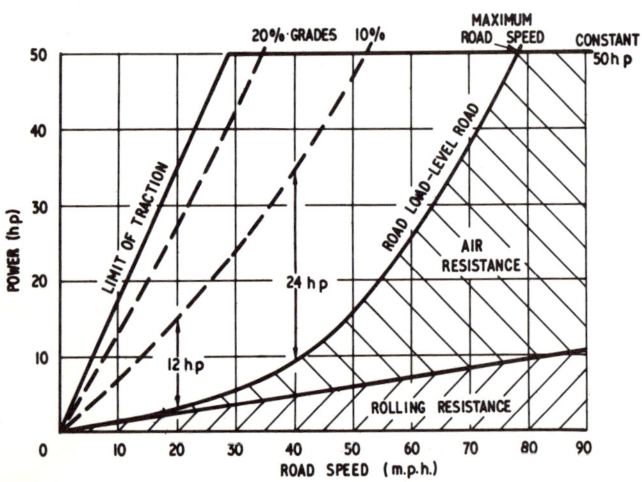

Fig. 1.6 Vehicle power requirements

on the graph as an area below a straight line. Air drag force increases with the square of air speed so that the energy absorbed increases very rapidly as the cube of the vehicle's speed. The graph shows that at higher road speeds most of the engine's power is being spent in overcoming these air resistances.

Engine power absorbed in hill climbing increases with the speed of the vehicle up a gradient and as shown on the graph, a vehicle which needs 12 hp for a 10% gradient at 20 m.p.h. will require 24 hp to ascend the same slope at 40 m.p.h. Thus, although the normal power requirement for ordinary road speeds is quite low, road vehicles need extra power for acceleration, hill climbing, according to road and traffic conditions. In general, most vehicles will require to have the full engine's power available at all normal forward speeds so as to be able to deal with any road conditions.

The use of full engine power at low vehicle speeds means very high thrusts or tractive effort, approaching the limit of traction, or grip, of tyres. Hence it is possible to fix a lower limit of the reduction ratio need for a vehicle. Fig. 1.6 shows a 'limit of traction' line which is positioned usually by considering a 25% grade. The area on the graph underneath this limit of traction and a horizontal line representing the maximum engine power (50 hp) represents the limit of power requirements for a vehicle. Since the vehicle does not often need less than the level road power, it is possible to consider the unshaded area bounded between the road load curve and the other two full lines as the basic power requirement for a vehicle. A transmission designer must provide suitable means so that the engine can operate satisfactorily anywhere between these maximum or minimum levels of power at each road speed.

1.5 RATIO REQUIREMENTS

Vehicle engines develop their power at high speeds of revolution, whereas the road wheels of a vehicle do not often exceed 1000 rev/min, and larger vehicles usually operate at much lower wheel speeds. A vehicle transmission, therefore, must include some reduction ratios in the drive line. The total required ratio is usually shared between an axle reduction and some gearbox or its equivalent. If a designer is concerned with a vehicle having a maximum speed of 80 m.p.h., equipped with 6 ft circumference road wheels and an engine which is to rotate at 5000 rev/min at that speed, then the reduction ratio is easily calculable.

$$80 \text{ m.p.h.} = \frac{88}{60} \times 80 \text{ ft/sec} = 118 \text{ ft/sec}$$

$$118 \text{ ft/sec} = 118 \times 60 \text{ ft/min} = 1180 \text{ rev/min of 6 ft circumference wheels}$$

$$\therefore \text{ Reduction ratio} = \frac{5000}{1180} = 4{\cdot}24 : 1$$

By considering some low gear, and requiring the vehicle to operate, say, on a 25% gradient, then the maximum engine speed might be developed at 27 m.p.h. By a similar calculation the total reduction ratio is 12·65. A typical choice of components for such a vehicle would be an axle reduction of 4·24, combined with a gearbox

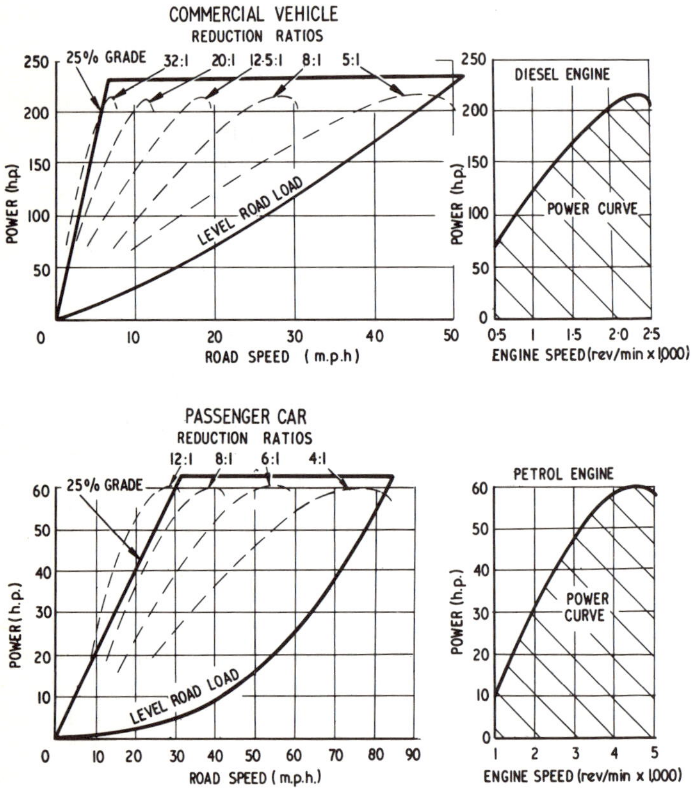

Fig. 1.7 Ratio selection for typical vehicles

having a low ratio of 3 : 1 reduction and a direct drive (1 : 1) top gear.

Operation of the vehicle and engine over the range of engine power and vehicle speed is usually achieved by providing a number of alternative ratios of connection between engine and wheels. Petrol and diesel engines used for motor vehicles produce their output over a range of speeds of rotation. Their maximum output is specified by a power curve which denotes the maximum power obtainable at each speed. By throttling the engine it is practicable to obtain a lower output at each speed. The transmission designer satisfies or 'fills' the required area on the power requirement graph by providing sufficient ratios of connection between engine and road wheels.

Fig. 1.7 shows diagrammatically how a number of reduction ratios are selected in order to satisfy the power requirements of a commercial vehicle and a passenger car. On the right of the diagrams are given the power curves of the respective engines. The area underneath the power curves represents the engine powers that can be obtainable at a given engine speed.

The diagrams on the left show this power curve when replotted a number of times to different reduction ratios between engine and driving wheels. In plotting these curves the highest 'gear' (i.e. smallest ratio) has been chosen so that a small reserve of power is available for acceleration when compared to the level road load curve. The lowest gear (largest reduction) has been similarly chosen to give a small reserve for the 25% grade. Ratios in between these two limits are spaced so as to cover the area in the best way possible.

It is not possible to cover the total requirement area completely and there are small gaps between gear ratios where less than full engine power is available. Different applications may warrant a greater number of available ratios, or an unequal spacing of ratios, but in general, 3–4 reduction ratios are satisfactory for motor cars, whereas at least 5 steps seem to be necessary for commercial vehicles. In these diagrams the road wheels are assumed to have a circumference of 6 ft in the case of cars, and 10 ft for commercial vehicles. In a later chapter it will be seen that operating economy can be improved by extra overdrive ratios which bring one or more power curves very close to the road load curve, so that there are advantages in having a wider spread of ratios than may be evident from these diagrams.

Vehicles require a reverse drive for manoeuvring, parking etc. and it is usual to provide a single reverse ratio to move the vehicle backwards. This ratio is similar in value to the lowest ratio so that the vehicle can reverse up steep gradients or under any difficult road condition.

1.6 OPERATING REQUIREMENTS

Having made the preliminary studies of the required layout, power and ratio requirements, a transmission designer can begin to form a mental picture of the transmission line and can even start to calculate the sizes of certain components. Before any design work can begin, the means of operating and controlling the transmission must also be considered. Some form of starting device is needed so that the vehicle can stop or start without stalling its engine which also needs to rotate at 800–1,000 rev/min before it will deliver power satisfactorily. Means for changing the overall reduction ratio whilst the vehicle is moving is required, preferably without needing a lot of skill from the driver. Driver's controls, gear levers or switches must be selected and positioned in the vehicle.

1.6.1 STARTING

Reference to Fig. 1.7 will show that all the power curves stop short at a low engine speed. There is apparently no power available for the car below about 8 m.p.h. and below 2 m.p.h. for the commercial vehicle. For starting the vehicle it is of course, necessary to speed up the engine and permit the transmission to 'slip' until the vehicle has accelerated to a speed when the particular reduction ratio can be properly connected.

Fluid drives are particularly useful for the starting operation because they have naturally smooth properties and can absorb a lot of the heat which is developed during the period of slip. A good heat capacity is particularly important in the case of commercial vehicles. Most passenger cars employ a simple friction clutch for starting since this finally connects the drive in a definite way and does not involve any 'slip' during normal running.

A wide variety of possible starting devices are available—some of which are described in Chapter 2—but the designer will be very much concerned with the size and cost of the unit as well as the operating characteristics when making his selection.

1.6.2 CHANGING RATIO

As a vehicle accelerates in a given transmission ratio, the engine speed rises and the driver will need to change to a 'higher' gear. This ratio change has, in the past, meant a gear change in a manually operated gearbox. By linkages connected to the gear lever, the driver can disconnect the drive from one pair of gears and connect it to another pair which, having differing numbers of gear teeth, give an alternative transmission ratio. Changing the drive from one pair of gears to another whilst the vehicle is moving and all transmission components are rotating, requires some practice and skill. There have been many attempts to simplify the procedure by the introduction of 'synchromesh' devices and other aids, up to and including fully automatic or variable transmission.

A designer can consider a wide choice of ratio-change methods ranging from the simple constant-mesh gearboxes still widely used for commercial vehicles, to the more modern automatic gearboxes which not only change gear without interrupting power, but also decide when the change should be made.

Again, the size and the cost of the device selected are important factors and must be offset against its performance and the type of vehicle considered.

1.6.3 ECONOMY AND UTILISATION

Some of the components in the transmission line are subject to fatigue and wear, and consequently may have a limited life. Clutch linings, and rubber components may need renewing and highly stressed gears and bearings may suffer from mechanical failure if the transmission is used beyond its intended life. In automobile applications it is usually necessary to draw a fine balance between the type, number and quality of components so as to achieve a minimum initial cost of a device consistent with the design life.

The initial cost consideration is usually a prime factor in passenger car design which is executed for a very competitive market. Effort to cut costs by reduced performance, however, are not usually successful. Three-speed gearboxes, therefore, are largely being displaced by 4-speed versions. Overdrive gears can be offered at the comparatively high cost of an 'extra' because this additional expense is justified by the benefits an overdrive brings.

Commercial vehicle transmissions are, of course, more expensive than their passenger car equivalents but the same compromise between life and initial cost is evident. It is difficult to achieve this compromise other than by the actual experience of a large number of vehicles operating over a wide range of duties and routes. Most large manufacturing units, however, employ test rigs to evaluate and compare differing designs by actually running gearboxes and components under special test conditions.

1.6.4 LOAD FACTOR

Vehicles are frequently judged by their maximum engine power, and maximum speed is another important design feature for motor cars, whereas maximum load is more important for commercial vehicles. In practice, however, the engine does not operate for very long at its maximum power condition. A quick way to prove this point is to collect information on the route speed and fuel consumption of different vehicles. The following table shows that most vehicles operate at well below half the maximum potential output of the engine. Load factor is the mean engine power expressed as a percentage of the maximum engine power.

Table 1.1 TYPICAL VEHICLE LOAD FACTORS

Vehicle	*Engine Power* (Max bhp)	*Route Average Speed* (m.p.h.)	*Route* (m.p.g.)	*Fuel Consumed* (gal/hr)	*Average Power* (bhp)	*Load Factor* (%)
Small Sports Car	50	45	35	1·3	15·6	31
Large Sports Car	150	60	20	3·0	36·0	24
Small Saloon	40	40	35	1·14	13·7	34
Large Saloon	120	50	18	2·77	33·3	28
3 Ton Truck	80	25	14	1·8	36·0	45
7 Ton Truck	120	25	10	2·5	50·0	41
London Bus	100	9	9	1·0	20·0	20

This table has been based on the assumption that 1 gal of petrol produces 12 bhp h from a typical engine whilst 1 gal of diesel produces 20 bhp h. Using these figures it is easy to determine the average rate of working for an engine. Traffic conditions are a controlling factor and it is very difficult to travel anywhere and consume fuel at more than 3 gal/h. Motorway travel can double this figure, perhaps, for some very powerful vehicles. Racing cars of 250 hp seldom consume more than 10 gal/h (120 hp average).

It will be clear from this table that although transmissions may be designed and tested for the maximum power condition, their utilisation in service will mainly be at much reduced power.

1.6.5 ROAD OPERATION

Vehicles travelling on any route are seldom able to maintain a fixed speed or engine condition. The vehicle will slow up on hills and for bends, and accelerate down gradients or to overtake other traffic. Not only will the engine speed be varying but the power output will change. On down gradients heavy vehicles will drive the engine which may be used partly as a brake by changing to a lower gear. On these over-run conditions, the transmission torques are reversed and gear teeth, bearings and couplings etc. must be considered and designed with regard to these over-run conditions.

Rough road surfaces, bumps and jerks experienced by the vehicle will also be communicated to the transmission so that the size and weight of a vehicle are nearly as important a consideration in transmission design as the engine output torque. Unfortunately, it is not possible to obtain information on actual operating experience until the complete vehicle is available in its final form, and a designer can only estimate for these effects by past experience of similar designs.

This chapter has touched on some of the major features of transmission design. As has been seen, there are many aspects to be considered in producing a completed design for a vehicle transmission. Transmission designers must in general seek a compromise between conflicting requirements and this is not easily done in a first appraisal. An initial design exercise can usually be improved by separate consideration of each requirement. Transmission design is very much affected by any changes in engine performance which may occur during the production life of a transmission, and also by the various different vehicle applications to which it may be applied. A successful transmission design is more likely to be one which, with due consideration to size, weight and cost, is capable of some improvement for every requirement, rather than one which has been designed too closely to an original set of design requirements.

BIBLIOGRAPHY

1. GILES, J. G., 'Transmission Ratio Requirements of Road Vehicles.' *M.I.R.A.*, Report 1957/4
2. BARTH, R., 'The Aerodynamics of Car Body Shapes.' *Engineer's Digest*, **17,** No. 10, 427 (Oct., 1956)
3. WHITE, R. G. S. and CARR, G., 'An Aerodynamic Survey of Vehicle Shape.' *M.I.R.A.*, Reports 1962/1; 1964/4; 1965/13; 1966/10
4. 'Automatic Transmission for Wolseley 18/85', *Automob. Engr.*, **57,** 124 (Apr. 1967); **57,** 344 (Sept. 1967)
5. GILES, J. G., *Automatic and Fluid Transmissions*, Odhams Press (1961)
6. 'The Ferguson Four-Wheel Drive for JENSEN CV8FF', *Automob. Engr.*, **56,** 232/3 (June, 1966)
7. BILLINGSLEY, W. F. *et al.*, 'The Rolling Resistance of Tyres as a Factor in Car Economy.' *S.A.E. Trans.*, **50,** 37–39 (Feb., 1942)
8. JOY, T. J. P., 'Tyres for High Performance Cars.' *S.A.E. Trans.*, **64,** 319 (1956)
9. CARR, G. W. and ROSE, M. J., 'Tyre Rolling Resistance Measurements on Road Dynamometer', *M.I.R.A.* Report 1967/2

10. THAW, D., 'Future Trends in Transmissions', *Comml. Mot.,* **124,** No. 3186, 14–21 (1966)
11. ANSDALE, R. F., 'Transmissions Survey', *Automob. Engr.,* **56,** 224 (June, 1966)
12. ANON., 'Hydrostatics—A New Way to go', *Steel,* 48 (Sept., 1967)

2

clutches and starting

Although gas turbines and other new engines are sometimes considered, road vehicles still mainly employ piston engines for their motive power. This engine develops its power over a range of speeds but needs to be rotating at a minimum speed before power can be taken from it. It is necessary therefore to have an uncoupling device so that the engine can be started and run up to this speed whilst the vehicle is stationary. In the past, a friction clutch has been used for this duty and a pedal is depressed to disconnect the drive. New types of engines or transmissions may some day do away with this clutch or coupling device but in general it is usual in most vehicles to find a friction clutch mounted on the engine flywheel to enable the engine to be connected and disconnected as required.

Having introduced this uncoupling device into the transmission line, there are certain fundamental problems which arise when the engine and transmission are to be connected again. Because the engine must run at a minimum speed of 800–1,000 rev/min before it can produce useful power then the vehicle must be moving at a given speed before the engine and transmission can be completely connected together. Study of the power curves seen previously in Fig. 1.6, for instance, shows that none of these power curves, even in very low gears, extends below speeds of 3–5 m.p.h. However low the gear ratio may be, there is still a minimum road speed at which the vehicle must be moving before any of the power curves can be utilised.

This chapter is concerned, therefore, with the starting of the vehicle, i.e. the movement of the vehicle from rest up to a low speed so that the transmission can be properly connected in a given ratio.

Similarly it is necessary to disconnect the transmission as the vehicle is coming to rest when the speed falls below that applicable to any transmission ratio. A friction clutch carries out these duties satisfactorily because during the starting operation it can transmit some engine power whilst slip is occurring, and this power is sufficient to accelerate the vehicle until the clutch plates stop slipping and the drive is rigidly connected. When coming to rest, the clutch pedal is depressed and the vehicle is stopped by its brakes, leaving the engine running at its idling speed. Although this 'transition' period from rest into motion is quite short, it is one of the more important features of the vehicle transmission requirement.

2.1 BASIC STARTING REQUIREMENTS

The requirements of any device intended to carry out the starting and stopping sequence of a vehicle can be specified under three main headings:

(1) To transmit engine torque and power smoothly during a period of 'slip' as the vehicle is moved from rest and until the engine is finally connected to the driving wheels (Starting Torque).

(2) To connect the transmission with minimum power loss for all normal driving (Efficiency).

(3) To disconnect the transmission cleanly without loss or 'drag' when required (Idling Drag)

These starting requirements are discussed in more detail below and with reference to particular starting devices which may have special properties which suit it for one of these requirements.

2.1.1 STARTING TORQUE

The first requirement usually presents most difficulty and Fig. 2.1 is useful in showing diagrammatically what the starting device has to achieve. This diagram shows that a car has to be travelling at 5–10 m.p.h. before the lower gears can be fully connected, and a 5-ton truck must be moving at 2–3 m.p.h. before its engine can produce useful power.

To start the vehicle it is necessary first to overcome the starting resistance which is quite small on a level road, but is much higher on a hill gradient. Considering Fig. 2.1a it is seen that in second gear, a typical car can provide a tractive effort or thrust of 250–

300 lb. This thrust is sufficient to achieve a start on a 1-in-10 hill but would provide no reserve for accelerating the vehicle. Second gear would, however, give a good start on level ground and leave a thrust of 225 lb available for acceleration which, initially, would be as high as 5 m.p.h./sec for a vehicle weighing 1 ton.

In starting this vehicle, the engine would be run up to a medium speed and then the clutch would be partially engaged and permitted

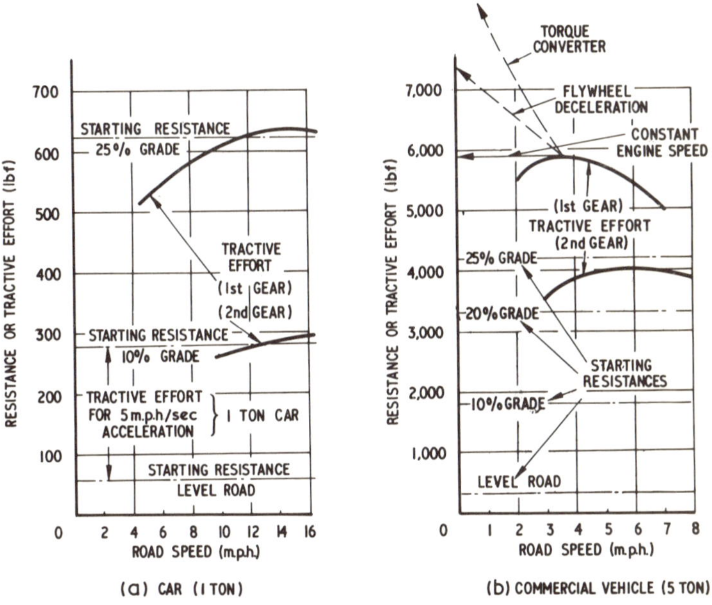

Fig. 2.1 Vehicle starting performance

to slip for about 2 sec until positive connection was achieved. The clutch would transmit a torque equal to engine torque during this transition period and this torque—converted to a tractive effort of 250–300 lb—would accelerate the vehicle to about 10 m.p.h. in this time. It will be seen that there must always be a good reserve of tractive effort available during starting so that the period of 'slip' is not too long. There is much heat generated in the clutch plates during starting and excessive slip will eventually destroy the friction material in the clutch.

First gear would be a much more satisfactory choice for starting up a hill because the available tractive effort is over 500 lb and the period of slip is reduced because the transmission can now be positively connected when the vehicle is doing only 5 m.p.h. On very steep hills it will be seen that this reserve for acceleration becomes less and Fig. 2.1a shows that on a 1-in-4 hill the engine can only just produce the torque necessary to move off up the hill. The period for which the clutch would have to slip might be excessive in this case.

Fortunately, however, it is possible to obtain some extra thrust for starting from the engine's flywheel. This effect is shown in Fig. 2.1b which is a similar diagram for a commercial vehicle. Resistances and tractive efforts are nearly ten times those for the car, and the transmission can be engaged at much lower road speeds of the vehicle. The transmission ratios of a truck are chosen so that there is more reserve of thrust for starting, and first gear has some extra effort available on any hill likely to be met in service.

Even so, it is useful to get the vehicle moving quickly, and with a heavier load to move, extra thrust is always helpful. Fig. 2.1b shows how some extra starting thrust can be obtained. Considering the arrowed lines which originate from the peak of the first gear tractive effort curve, then it is seen that if the engine speed is kept steady at its highest torque, then the slipping clutch transfers the equivalent tractive (6,000 lbf) effort during the starting period of 0–2 m.p.h.

For a steep hill, it is possible to speed up the engine to near its maximum and to apply the clutch rather less gently so that the transmission gains some additional torque whilst the engine and flywheel are being slowed down to the engagement speed. Most clutches are designed to carry at least twice engine torque before slipping (clutch pedal released) so that it is theoretically possible to double the starting torque by this means. This procedure would give a fierce start and the energy stored in the engine and flywheel would last only for a short time. Firm treatment of the clutch, therefore, can provide some extra torque over the starting period and a possible initial torque curve is shown on the figure as 'flywheel deceleration'.

A third curve is shown on the diagram—marked 'Converter'—and this shows how even greater starting torque can be obtained by means of a fluid converter. This type of fluid drive is described in later pages, and various designs are produced but all are designed to convert the engine's power so as to give high starting torques. The hydrokinetic fluid converter, as it is more correctly called, can be fitted in place of the friction clutch, and is usually employed with a number of gear ratios. The converter, however, does not disconnect the engine completely for gear changes and some other form of clutch or coupling is required for this purpose.

2.1.1.1 *Flywheel deceleration*

A vehicle's engine and flywheel have considerable rotary inertia, and because of this there is an amount of energy stored in the rotating parts. If the engine speed is reduced, as is normally usual, during the clutch engagement, then this energy is available as extra torque which increases that already being developed by the engine. It is possible to calculate this starting torque as follows:

$$\text{Starting torque} = t_e + I_e\alpha \qquad (2.1)$$

where I_e is the total engine inertia (lb ft sec^2) α is the rotary deceleration

(rad/sec^2) t_e is the engine output torque (lb ft)

In practical terms I_e would range from 0·1–0·4 for cars and up to 1·0 or more for commercial vehicles (lb ft sec^2). Hence if the engine is engaged at a high speed and loses 2,000 rev/min during a sharp (1 sec) engagement ($\alpha = 209$) the $I_e\alpha$ term could be

$0{\cdot}4 \times 209 = 84$ lb ft (cars)
$1{\cdot}0 \times 209 = 209$ lb ft (commercial vehicles)

It will be seen that in approximate terms, there is prospect of doubling the engine's torque for starting by use of a fierce clutch engagement, and only by limiting the maximum torque transmitting capacity is it possible to prevent even higher torques being developed during a jerky start.

It will be noted that this extra torque is available only for the short time (1 sec) of the 'slip period' and the energy stored in the flywheel is practically expended during this short period of extra working.

2.1.2 EFFICIENCY

A simple friction clutch has good properties for vehicle starting, especially, as seen in the previous paragraphs, when it is possible to transmit temporarily a higher torque obtained whilst the engine and flywheel are decelerating. The torque developed on the driving and driven plates of a clutch must always be equal but opposite in sign so that the efficiency of a clutch rises as the output speed rises (see Fig. 2.2). Once a friction clutch is engaged, there is no slip between the clutch plates and there is no loss of power in this form of coupling.

Fig. 2.2 shows a view of a typical friction clutch which is mounted on an engine flywheel. A clutch plate consisting of a thin disc flexibly mounted on a central boss which takes the drive to a shaft—usually the gearbox shaft. The clutch plate is lined on both sides at its outer diameters with an asbestos type of friction material. The friction material is clamped between the rear face of the

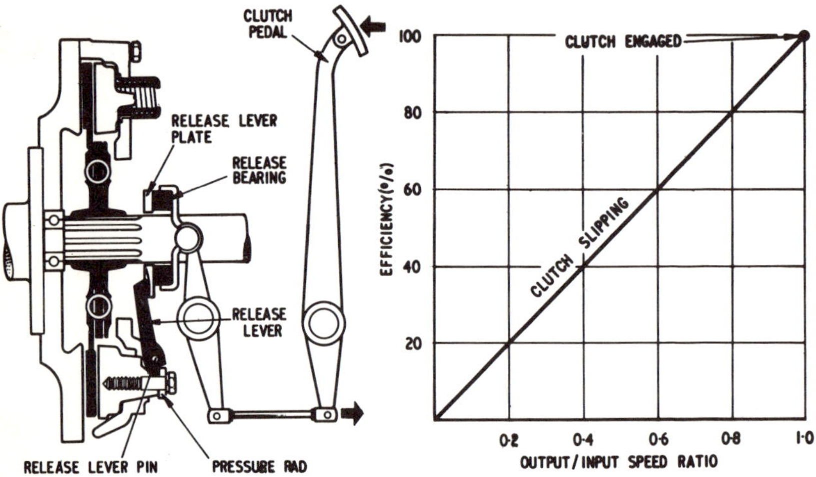

Fig. 2.2 Basic elements and coupling efficiency of a friction clutch

flywheel and a pressure plate which is loaded by a series of springs mounted in a clutch casing bolted to the flywheel. Under normal driving, therefore, the clutch plate is loaded between two metal faces by a total load equal to the total spring load. There may be a dozen springs each exerting a load of 100–200 lb, so that the total spring load may be nearly 1 ton.

The grip of the clutch plate is dependent on the coefficient of friction of the friction material and there is a good range of possible coefficients available to a designer.

2.1.2.1 *Torque capacity*

The torque rating of a simple flat plate clutch is determined by the friction force μF acting at some mean effective radius R, so that clutch torque may be expressed

$$\text{Clutch torque } t_e \text{ (lb ft)} = \mu FR \qquad (2.2)$$

where μ is coefficient of friction, F is the total clutchload (lb) and R is the mean effective radius of friction area.

As has been mentioned previously, F is likely to be 2,000 lb or more and R will be from 0·3 to 0·6 ft. The mean effective radius may require more accurate calculation depending on the pressure distribution of the clutch load across the area of the friction material but cannot be too far from the geometric mean radius which gives an approximate clutch rating. A considerable range of coefficients are available for clutch friction materials which are often about 0·3–0·4 but can be nearly double this figure.

In general, however, it is more important to choose a material with good properties at high temperatures and it is even more important to have a large enough area of friction material to absorb the heat generated. Manufacturers may recommend materials by means of engine torque or horse power per square inch of material. Torque transmission requirements can be achieved by adjustment of spring load, and a practical slip test is usually employed to ensure that the clutch will transmit about twice engine torque before slip occurs. A single plate clutch of this type operates 'dry'. Oil may leak into the clutch compartment but small amounts are rapidly evaporated by the heat generated during slipping.

The clutch can be disconnected by levers between the pressure plate and the clutch casing and attached to a release bearing or thrust block, usually made of carbon, although a thrust block containing metal balls or rollers is sometimes employed. The clutch pedal acts on this thrust block or release bearing and, if sufficient load and leverage is employed, compresses the springs and withdraws the pressure plate, thus disconnecting the clutch.

The mechanical arrangement of this simple friction clutch can be seen from this diagram and description to be such that no losses are incurred when the clutch is engaged—except perhaps for slight windage losses due to the rotation of the clutch casing in air. It is difficult to visualise any starting device which has a higher efficiency when coupled than a friction clutch of this type.

2.1.3 IDLING DRAG

Similarly, when the friction clutch is disengaged by depressing the pedal, the clutch plate is completely freed from the engine and the shaft will stop. There will be a small amount of loss in the thrust bearing which is loaded only when this pedal is depressed. This loss will show up, for instance, as a slight decrease in engine idling speed, or as a small increase in fuel consumed in a governed diesel engine. For most practical purposes, however, it may be said that a friction clutch has no idling drag.

The same cannot be said, however, of a fluid coupling or fluid flywheel as it was originally called. This drive is often employed on buses and for other applications where frequent starts are necessary. The coupling consists of a driving and a driven member which are immersed in fluid contained in a casing which usually replaces or forms part of the engine flywheel. Hence the use of the term 'fluid flywheel', which has now largely fallen into disuse because there are a large variety of different fluid devices now in use.

The fluid coupling operates by virtue of the fluid which is ejected out of the impeller (the driving member or pump) and is directed on to vanes or blades in the turbine (the runner, or driven member). This circulation of fluid exists all the time that there is a difference in the speeds of rotation of the impeller and turbine. Hence, when the vehicle is stopped, and its engine is idling, there is still a noticeable transmission of torque through the coupling, and of course the basic idea of the coupling is to exert torque from a speed difference between the two members. Efforts to reduce idling drag tend to reduce the torque capacity of the coupling and hence its efficiency when driving in the normal condition (see Fig. 2.3). Various improvements in design have reduced this idling drag, but the only way that a fluid coupling can be completely disconnected is by emptying it of fluid.

A typical fluid coupling for automotive purposes is shown in Fig. 2.3, where it will be seen that the impeller forms part of the casing and it is bolted on to the engine flywheel. The turbine (runner) fits inside this casing and drives an output shaft which extends from the centre of the impeller casing. The drive is filled with fluid (apart from a small expansion volume) and an output torque exists as soon as the impeller is rotated. The circulation of fluid is generated by flat radial vanes in the impeller and turbine. Its efficiency during the starting of a vehicle is proportional to the speed ratio and is similar to the friction clutch, except that it never completely reaches the 100% (locked up) condition of the clutch.

Fluid drives, such as the coupling or converter follow the relationship:

$$T = CN^2D^5 \tag{2.3}$$

where T = transmitted torque, C = a constant, N = Speed, D = Diameter.

It will be seen that a small change in diameter gives a large change in the transmitted torque, and also that the torque developed on the output shaft when it is held stationary varies as the square of the input speed.

Thus if a coupling is designed to transmit 100 lb ft at 1,000 rev/min input it will transmit over 16 lb ft at 400 rev/min. This 16 lb ft represents an idling drag which may cause the vehicle to

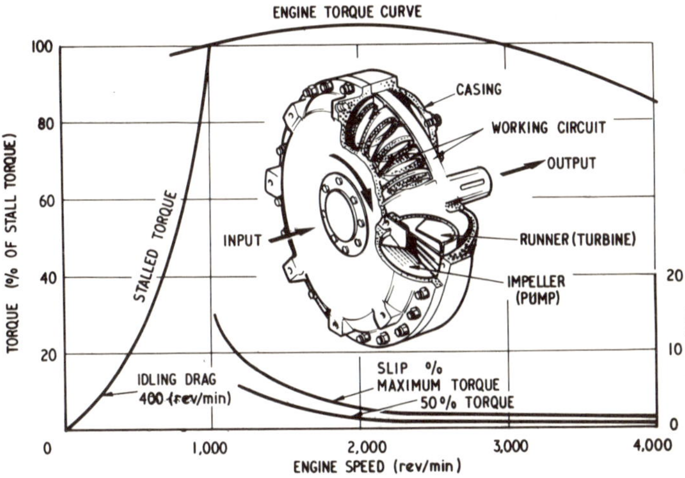

Fig. 2.3 Sectional diagram and performance data of a fluid coupling

move if the brakes are not applied, or alternatively introduces some drag on the engine which may thus consume extra fuel.

The advantages of a fluid coupling, however, are seen in its extremely smooth operation for starting and, also, in its capacity for absorbing heat during difficult or frequent starting. It can be shown that half of the engine power is wasted in heat during the period of engagement of a clutch or coupling from rest. In large vehicles and engines, this amount of heat can result in excessively high temperatures. A fluid coupling containing a quantity of fluid has a considerable heat capacity, and provides a large surface area to dissipate heat to air. Hence, it can tolerate severe use, possibly misuse, better than a friction clutch which may deteriorate rapidly after attaining excessive temperatures. Moreover, there is little wear in a fluid coupling which apart from occasional replenishment of fluid, can be expected to last indefinitely.

As compared to a friction clutch, therefore, the fluid coupling is less satisfactory because of its idling drag and also for some inefficiency because there is still some slip of about 1–2% even at maximum speed. The fluid coupling is more satisfactory for life and wear—it needs no re-lining etc.—and can be relied upon to give smooth starts, free from jerks or judder.

2.2 SPECIAL CHARACTERISTICS

In addition to the three basic requirements, various starting devices have no other properties or characteristics which particularly suit them for the starting operation. Some of these properties are discussed below.

2.2.1 TORQUE MULTIPLICATION

The excellent properties of the fluid coupling during the starting, or engagement period are surpassed by another type of fluid drive, called a 'converter' or 'converter-coupling'. The converter-coupling has been widely used in automatic transmissions and operates on the same principles as the fluid coupling. The converter versions contain at least three rotary elements usually known as the pump (impeller), turbine and reaction member. The reaction member is connected to an external casing to provide a reaction on the circulating fluid.

Whereas in a simple two-element coupling the input and output torques are equal (but opposite), the introduction of a third reaction element enables the turbine to develop a higher output torque than input. The difference between input and output torques is provided by the reaction member. With a three-element converter (Plates 2.1 and 2.2 show typical designs), it is possible to obtain a torque multiplication during the starting operation so as to give a more rapid acceleration and a reduced engagement period. Three-element fluid devices tend to be less efficient than the two-element coupling for normal driving and designs giving high torque multiplication ratios tend to give a further loss of efficiency. Thus although a torque multiplication of 4 : 1 has been achieved, it is more usual for converters to double the input torque for starting (see Fig. 2.4). The reaction member is usually mounted on a free-wheel device so that when starting has been completed, the reactor 'freewheels'. In this way the converter reverts to the simple coupling operation for most of its operation. Hence, the term converter-coupling.

Converter-couplings introduce further idling-drag because of their torque multiplication properties. When employed in vehicles it is necessary to make some special provision for this drag. In most cases the transmission will be disconnected completely by some other form of coupling for a 'neutral' position. The Porsche 'Sportomatic' transmission (see Fig. 2.5) is a typical example of this combination. Used with a friction clutch it is possible to disconnect the drive temporarily to remove idling–drag.

For other versions when no additional clutch is employed, and

Plate 2.1 Basic elements of a hydrokinetic converter-coupling left–turbine; centre–reactor; right–pump. (By courtesy of Borg-Warner Ltd.)

Plate 2.2 Basic elements of a hydrokinetic converter-coupling manufactured from light alloy castings. (By courtesy of J. Brockhouse and Co. Ltd.)

when a gear is engaged ready for starting it may be necessary to apply a brake to the vehicle to prevent movement of the vehicle due to this idling drag.

At this point, therefore, it is possible to state that although fluid drives provide excellent automatic starting and fulfil the first requirement of the starting device in a superior manner, they lack the efficiency for normal driving and the clean disconnection of the mechanical friction clutch. All the devices described may be employed on vehicles depending on the relative standard of

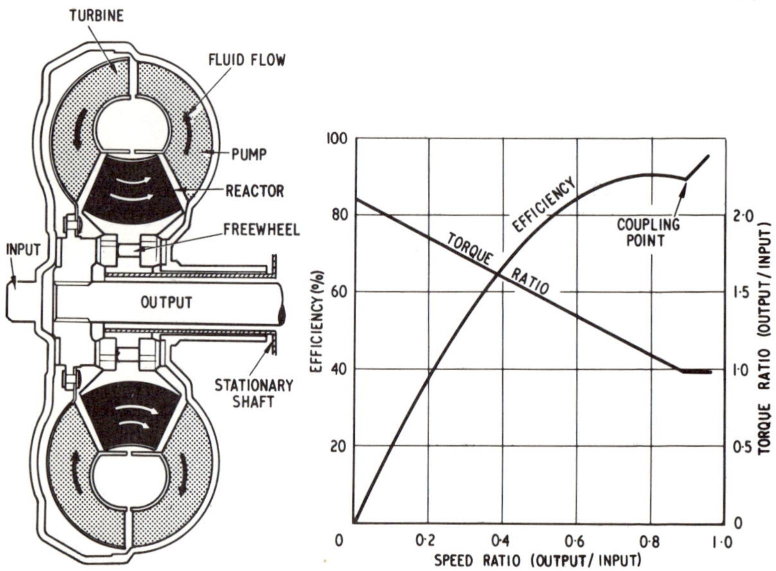

Fig. 2.4 Typical operating performance of a hydrokinetic converter-coupling

performance required, and bearing in mind their comparative initial and operating costs. Some further aspects of performance are now considered as a means of selecting between the alternative mechanical and fluid equivalents.

2.2.2 FLYWHEEL ENERGY

As shown briefly on Fig. 2.1b it is possible to obtain some additional starting torque from energy which is stored in the engine and flywheel. This additional torque becomes available when the engine speed is slowed during a clutch engagement process. When rotating at speed a flywheel contains kinetic energy due to its rotation and when a friction clutch is engaged for starting, this clutch must exert not only the engine torque but also some extra torque to slow down this flywheel.

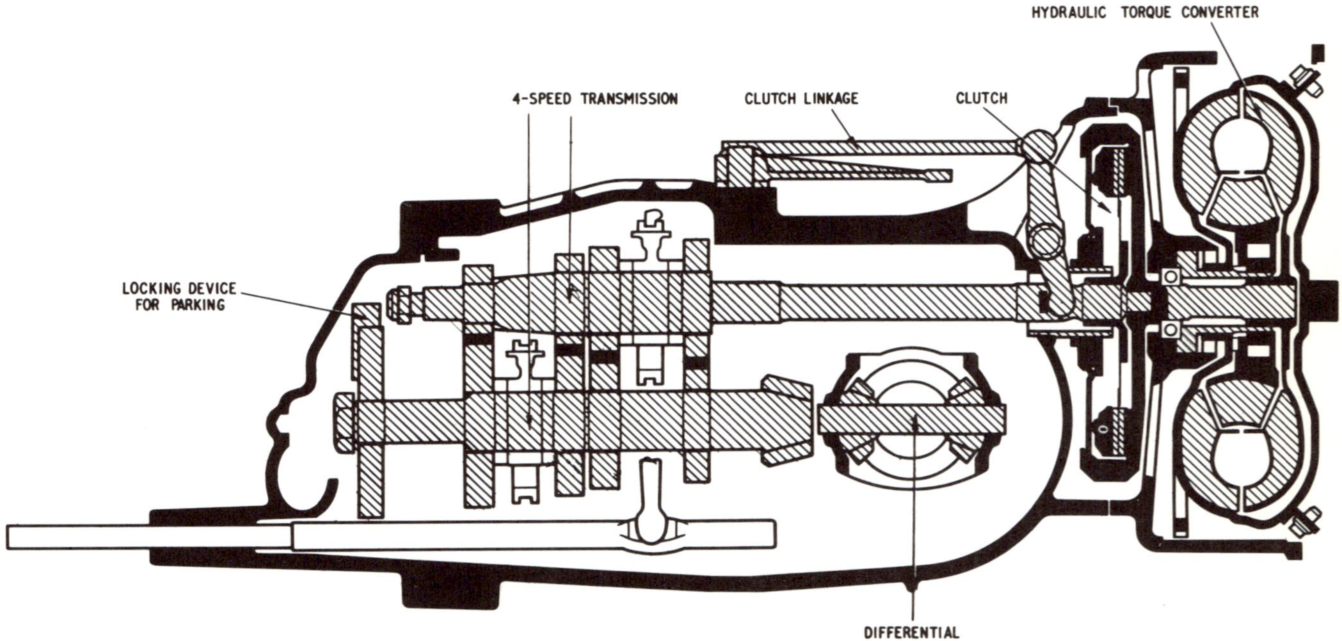

Fig. 2.5   Fluid drive combined with a friction clutch in the Porsche 'Sportomatic' transmission

Extra torque, therefore, is available in any vehicle start (or other manoeuvre) which involves a deceleration of the engine and flywheel. Most normal starts with vehicles having a conventional friction clutch introduce such extra starting torque probably without the driver realising the fact. Experienced drivers may deliberately employ higher engine speeds for fast, or difficult starts in order to make use of this available starting torque.

Automatic clutches, or transmissions, which may hold the engine speed constant during the starting sequence cannot make use of such extra torque. This feature is one of those which have contributed to the sluggish feeling which some automatic devices produce when employed on vehicles for starting.

The engines and flywheels of commercial vehicles are usually heavier than those for passenger cars but in any case the limit of extra torque is dependent on the maximum torque of the clutch or coupling used for starting. The size of the flywheel determines for how long a period the extra torque can be maintained, and a larger flywheel inertia will extend the period for which extra torque is available. Cars can expect to obtain 100% extra torque for about one second and vehicles with heavy diesel engines would expect to maintain this extra torque for 1·5–2 sec.

From theoretical consideration, this extra torque would be maintained for a short period and would then drop, on engagement, to the normal engine torque. In practice, the friction properties of a clutch would smooth this jerk, but under maximum conditions, the friction clutch start would be rather fierce. Fig. 2.6 shows diagrammatically the starting torques of a friction clutch compared with a fluid converter-coupling giving roughly 2:1 torque multiplication. Clearly the converter will maintain its extra torque for a much longer period of time and if the throttle is kept open will provide extra torque almost up to the engine's maximum torque speed. It will be noted, however, that the converter's multiplication begins to fall as soon as the vehicle begins to move, because this type of fluid drive quickly reverts to its coupling operation.

2.2.3 TORQUE LIMITATION

Although the design of axles, gearboxes and other transmission components is normally related to the maximum torque produced by the engine, it has been seen that by use of energy stored in a vehicle's engine and flywheel, considerably higher torques can be developed for short periods of time. Similarly the large inertia of the vehicle when moving at speed can develop dangerously high torques in the transmission if a clutch or coupling is engaged fiercely when there is a speed difference between the coupling members.

The starting clutch (and any other coupling in the transmission line) can act as a safety valve if it can be designed to slip at some limiting torque value. Hence, although it is useful for starting to

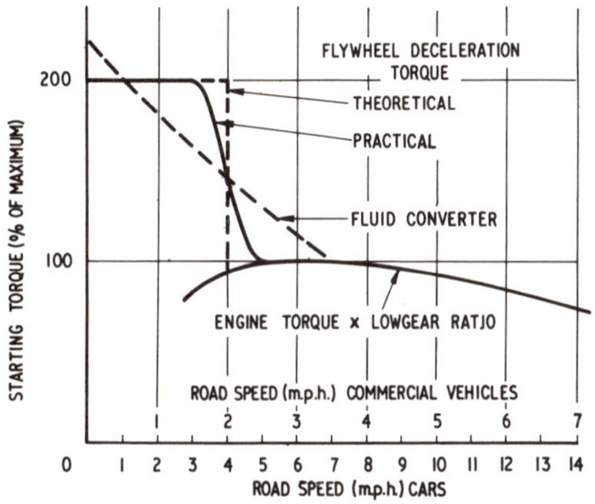

Fig. 2.6 Additional starting torque—comparison of the starting torque—flywheel deceleration versus fluid converter (maximum condition)

be able to develop extra torque from the flywheel, it is dangerous to employ clutch springs loaded too highly, because the transmission will have been designed to accept much lower torques.

Consideration of any form of starting device should include a study of its torque limiting properties, especially if it is intended to replace a coupling which is, effectively, the transmission's torque limiter or 'safety valve'.

2.3 MISCELLANEOUS STARTING DEVICES

2.3.1 OTHER FORMS OF FRICTION CLUTCH

The simple single plate friction clutch is so close to ideal for starting vehicles that improvements in smoothness and life have been sought by alternative designs. In some cases, the conversion of the single plate (dry) clutch to a wet version, by sealing the clutch compartment and providing a supply of oil, has given good improvement. The lubricated friction surfaces require much heavier loads to transmit equivalent torques, but the presence of oil or fluid reduces wear and plate temperatures. The supply of fluid also adds to the heat dissipation properties of the assembly.

Not many examples of the single plate wet clutch are available, but the main difference from a 'dry' version is the need to obtain extra clamping force so as to give the required transmitted torque on account of much lower friction coefficients of the wet plates. Hence, wet-plate clutches are usually seen in automatic forms where external hydraulic power can be introduced to exert the necessary clamping forces.

A cone clutch presents one way of obtaining higher transmitted torques for a given clamping load. In this design which is seen in both wet and dry versions, the angle of the cone shaped friction faces is such that a multiplication of 4 or 5 : 1 is obtained in this load. If θ is the included angle of the friction cone, it can be shown that for a clutch normal load F the clutch torque becomes

$$t_c = \mu F R \cot(\theta/2) \qquad (2.4)$$

and for a 30° cone angle the multiplication factor is 3·7. Fig. 2.7 shows a well known example of a cone clutch, which is employed

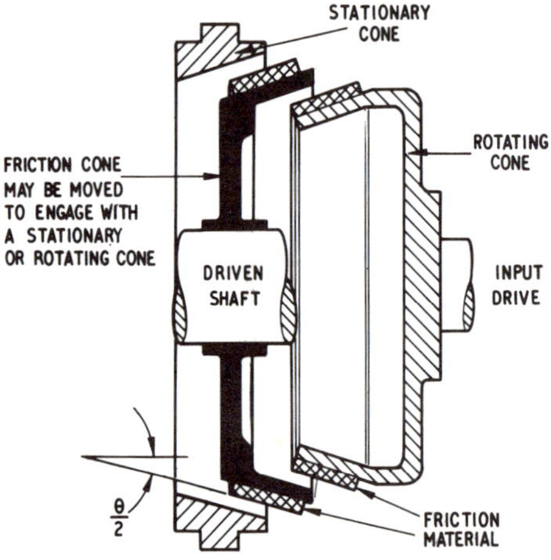

Fig. 2.7 Diagram of a 2-way cone clutch. (Laycock)

in an overdrive gear. This interesting version has friction material on both inside and outside faces of the cone and the moving cone acts as the male member of the brake which holds the cone stationary and as a female member of a clutch which drives the cone at input speed. Similar cone clutches have been employed

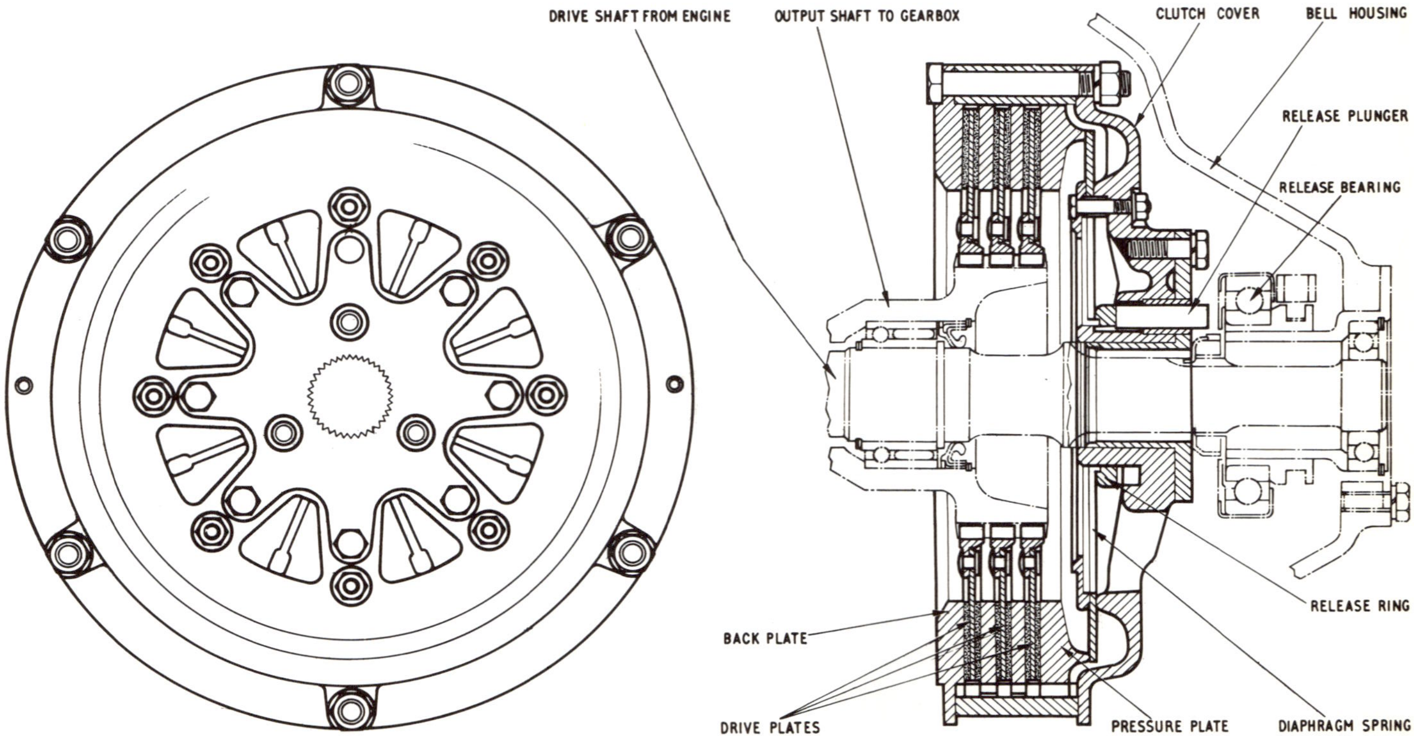

Fig. 2.8 Diagrams of a multiplate friction clutch. (Borg & Beck)

previously for vehicle starting and may be seen again in the future.

A multiplate clutch is a further common type of friction clutch. Used wet or dry, it contains a stack of friction plates connected alternately at the inner and outer diameters. The object of this arrangement is to increase the life by spreading the load and thermal effects over a large area of friction material. This type of clutch (see Fig. 2.8 and also Plate 2.3) is frequently employed for arduous duty because it has a good thermal capacity to absorb the large amounts of heat generated.

Multiplate dry clutches are employed for larger commercial vehicles and for transmitting very high torques in a small space

Plate 2.3 Exploded view of a multiplate 'dry' friction clutch. (By courtesy of Borg and Beck Co. Ltd.)

(racing cars for instance). Multiplate wet clutches are usually hydraulically controlled and operated, when the hydraulic fluid is used for cooling and for applying the clamping force by hydraulic pressure. In these latter versions, alternate clutch plates are frequently made of dissimilar metals and the application of hydraulic pressure for clamping is carefully regulated and controlled to give torque transmission characteristics suitable for changing from one gear ratio to another.

A device which is gaining in application is the diaphragm spring (Fig. 2.9a and b). This spring can be employed in most forms of friction clutch either to apply the clamping force needed in single plate clutch or to release the plates in a hydraulically operated friction clutch. The diaphragm spring consists of a dished circular plate, which is stamped in the form shown. The spring is usually supported at the outer edge and as pressure is applied to the centre. The spring may be set initially to a given shape so that some of the applied load is obtained during assembly and the operating ring or plate requires only a very light load to apply or release the clamping force.

2.3.2 AUTOMATIC CLUTCHES

Fluid drives are superior to the simple friction clutch for automatic use because they remove the need for a clutch pedal and automatically take up the drive when the engine speed increases.

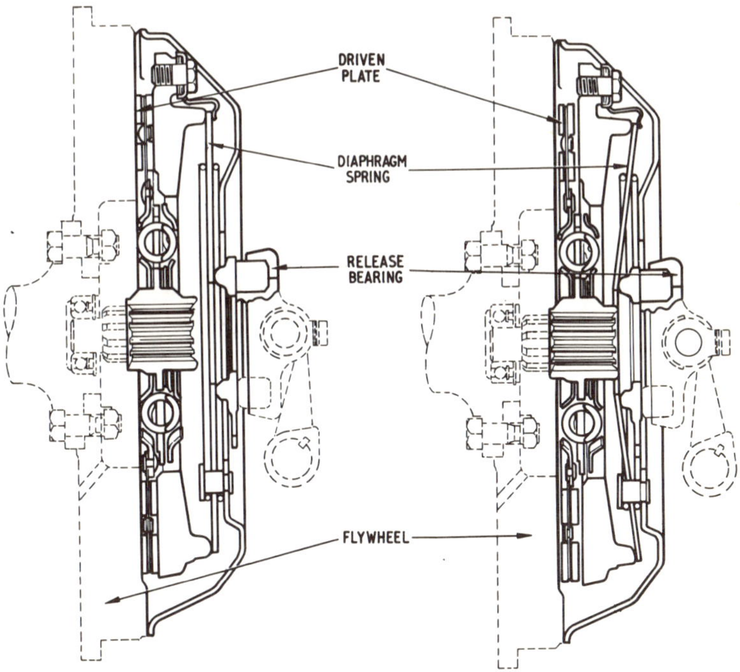

Fig. 2.9a Diaphragm spring in a friction plate clutch

Many efforts have been made to obtain automatic operation of a friction clutch which will commence to engage after the engine has reached a given speed. These devices have been based on hydraulic, pneumatic, or centrifugally operated mechanisms, and there are various combinations of these forms. These systems are too

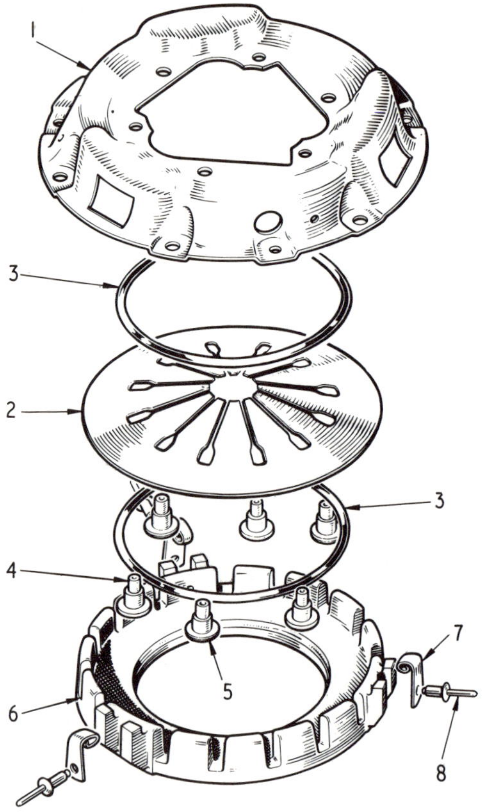

Fig. 2.9b Exploded view of the diaphragm spring: (1) Clutch cover; (2) Diaphragm spring; (3) Fulcrum rings; (4, 5) Locating studs; (6) Pressure plate; (7, 8) Clips. (Automotive Products Ltd.)

numerous to mention individually but the basic principles of each will be described briefly.

Centrifugal devices have mainly been attached to the flywheel so that bobweights could exert the necessary clamping force on the friction plates or shoes to give a drive as soon as a given engine speed has been reached. One of the simpler devices is seen on DAF vehicles, which has a number of spring mounted friction shoes attached to the flywheel (see Fig. 2.10). These shoes rotate inside a cylindrical drum, rather like a normal brake drum, which drives the gearbox shaft. As the engine speed rises these friction shoes fly outward exerting pressure on the driving drum. As soon as the centrifugal pressure is sufficient to overcome the starting torque, the driving drum starts to rotate. With a further increase in speed

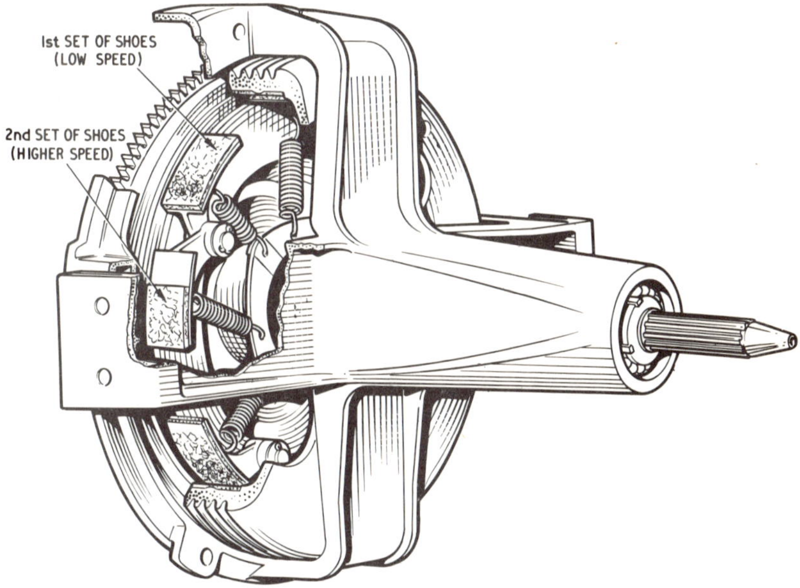

Fig. 2.10 Centrifugally operated friction clutch. (DAF)

a second group of friction shoes come into operation so that there is sufficient friction to transmit full engine torque. Unfortunately it is not easy to disconnect the drive once the vehicle is in motion and this simple type centrifugal clutch needs a further clutch or coupling to permit gearchanges to be made, although in the DAF cars no such gearchanges are necessary.

In another centrifugal clutch a comparatively small change to a normal friction clutch assembly permitted automatic starting. In this case centrifugal bobweights were introduced on the clutch housing so that the normal clutch springs were not depressed until centrifugal force on the bobweights was sufficient to overcome their spring load. Stops on the bobweights limited their movement so that at a high engine speed the clutch springs were depressed to a typical clutch load and the clutch pedal could be operated in a normal manner for gear changes (see Fig. 2.11).

Pneumatic devices operate by employing engine manifold vacuum pressures to act on a piston and move the clutch levers.

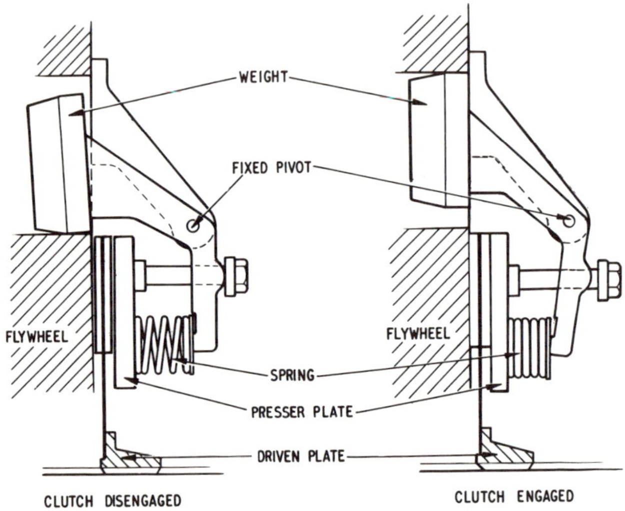

Fig. 2.11 Diagrams illustrating a centrifugally operated friction clutch which could also be disengaged by external means when rotating

One of these devices employed this vacuum pressure to depress a piston which acted on the clutch lever and depressed the clutch. As soon as the engine started the clutch was disengaged. When the engine speed was increased a solenoid operated by the electrical generator was employed to open an air bleed so that the clutch springs overcame the vacuum force and engaged the clutch steadily so as to move the vehicle.

By breaking this electrical solenoid circuit—using a gear lever switch perhaps—it was possible to close the air bleed and disengage the clutch for gearchanges. As a result a normal vehicle could be changed into two-pedal control merely by the addition of this vacuum cylinder with electrical solenoid control.

Hydraulic clutches may employ hydraulic pressure, available from some external source, to operate the clutch levers direct as in the case of the Citroen DS 19, or alternatively to operate directly on to the clutch plates to provide the clamping force. In either case the clutch and hydraulic system are usually part of some more complex automatic system and will be considered more fully in later chapters.

Although the centrifugal and pneumatic clutches gave satisfactory performance for starting they are in general rather too simple to obtain smooth and acceptable gearchanges under widely varying conditions of load and speed.

2.3.3 ELECTRICAL AND MAGNETIC COUPLINGS

The powerful forces that are obtainable by magnetism have frequently been employed for clutches, couplings and brakes. Electric currents flowing in armature coils can be employed to develop large forces which, in early attempts, were employed to pull a clutch driven plate into contact with a driving armature mounted on the flywheel. The friction drive was obtained by the friction between the armature core and the driven plate. Unfortunately, the materials most suitable for generating high magnetic fluxes are not usually very good for wear properties when sliding, especially when high temperatures are developed.

Consequently versions such as the Ferlec and Gravina magnetic clutches were designed so that the magnetic system merely provided a clamping force and both designs relied on normal friction'material for the driven plate. Magnetic clutches of this type rely on an electrical supply, and there are sometimes difficulties with rotating contacts (slip rings) which take the current to armatures revolving on the flywheel. Disappearance of the vehicle's drive due to some electrical failure is regarded unfavourably by some engineers, but a vehicle is nearly always immobilised by a serious electrical fault regardless of transmission design.

Another type of electrical coupling is known as the 'magnetic powder' coupling. In this type, which is illustrated in Fig. 2.12 and Plate 2.4, the magnetic forces are developed by means of electric windings, armatures etc. as in other types but containing air gaps which can be filled with iron powder when the coils are energised. The coils and armature may rotate on the flywheel, when slip ring contacts are required, but other versions have shown that a stationary assembly can be employed to couple two rotating members. The magnetic circuit is arranged so that loose iron dust of carefully prepared particle size will pack into spaces between the driving and driven members of the coupling.

Under conditions of 'slip', friction forces are developed by the forces acting on the iron particles as a result of the magnetic flux. One virtue of this arrangement is that the friction coefficient does not appear to be affected by the 'stick-slip' variations of normal friction faces. The transmitted torque of a coupling is dependent mainly on the applied electrical current. Suitable control of the magnetising current may be employed to obtain a required sequence of output torque for starting and gearchanges etc.

2.3.4 HYDROSTATIC COUPLING

The growing use of hydraulic equipment is seen in many fields of engineering. Items such as power-steering, power brakes etc. are

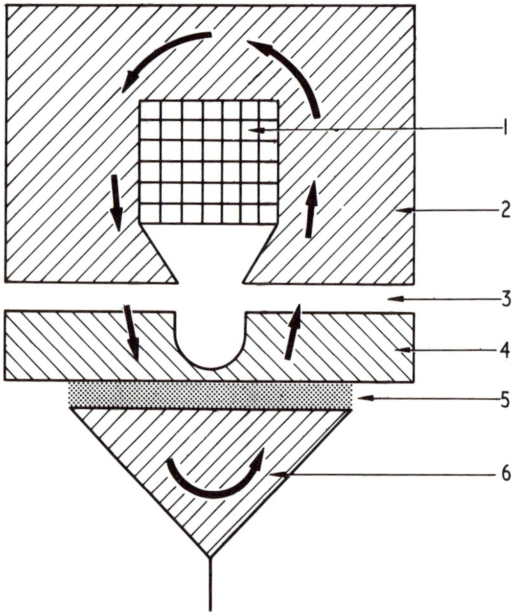

Fig. 2.12 Schematic drawing of a 2-element magnetic powder coupling employed in an automatic transmission and showing the principles of the powder gap: (1) Coil; (2) Stationary field; (3) Air gap; (4) Driving member; (5) Powder gap; (6) Driven member. (Smith Industries Ltd.)

readily available in automobiles and are examples of the use of hydraulic fluid employed at pressure to produce power. Automatic transmissions regularly use an auxiliary hydraulic circuit for the operation of clutches, brakes and other transmission components.

Vehicle transmissions based on the use of pressure hydraulics in various forms have been projected, developed and are in use for various specialised applications, and may soon spread to more common uses.

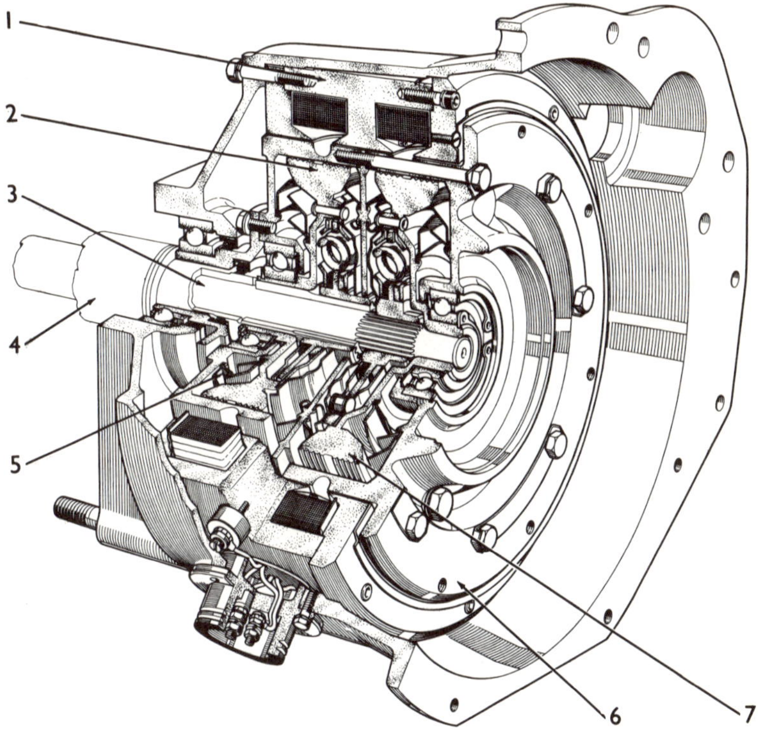

Plate 2.4 Cutaway drawing of a two-element magnetic powder coupling. (1) Stator (2) Driven member (indirect) (3) Gearbox main shaft (4) Gearbox outer shaft (5) Labyrinth (6) Driving member (7) Driven member (direct). (By courtesy of Smith's Motor Accessories Ltd.)

One of the advantages of the various forms of hydrostatic transmissions is the fact that coupling or decoupling is easily effected by means of a by-pass valve which permits fluid to circulate freely when this valve is open. This simple valve effectively replaces the clutch or other coupling required with other forms of transmission (Plate 2.5).

Problems of starting torque, efficiency and idling drag will, however, arise in the design of this by-pass valve. The flow characteristics of the valve will be selected by studies of starting torque and drag requirements, whereas the efficiency of coupling is dependent on leakage loss of the valve, which should be negligible.

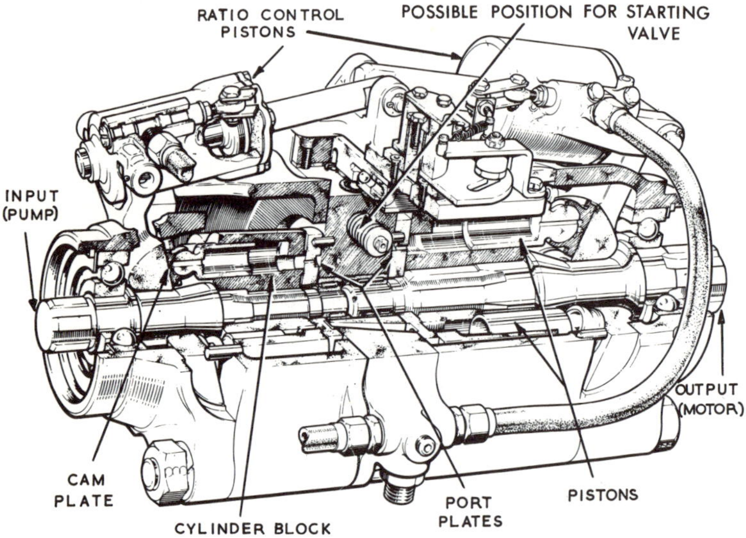

Plate 2.5 The starting properties of a hydrostatic transmission could be obtained by means of a simple valve inserted between the main fluid ports. (By courtesy of Joseph Lucas Ltd.)

By suitable control of the ratio control pistons, any required starting torque characteristic could be obtained from the transmission which, being a fluid drive, has very adequate heat dissipation properties.

SUMMARY

The starting duty for vehicles is characterised by a difficult period of torque transmission under the conditions of slip for which hydraulic and fluid devices are best suited because of their heat absorption or dissipation properties. The efficient connection and clean disconnection requirements are best satisfied by mechanical clutches, and most other starting devices tend to make some compromise between these basic requirements. The ideal solution has probably yet to be found and as seen above may come in the form of some transmission which does away with the need for a separate starting device.

BIBLIOGRAPHY

1. SINCLAIR, H., 'Recent Developments in Hydraulic Couplings', *Proc. Inst. Mech. Engrs. (London), Automob. Div.*, **130,** 175 (1935)
2. GILES, J. G., 'A Review of Hydrokinetic Fluid Drives', *Proc. Inst. Mech. Engrs. (London), Automob. Div.*, 43 (1956/7)
3. ANON, 'The Ferlec Clutch', *Automob. Engr.*, **45,** 81 (Feb., 1955)
4. ASHAUER, K. and HARMUTCH, H., 'The Selector—Automatic Transmission of the VW 1500', *A.T.Z.*, **69,** 288–293 (Sept., 1967)
5. ANON, 'Smiths Automated Transmission', *Automob. Engr.*, **56,** 359 (Sept., 1966)
6. AUTOMOTIVE PRODUCTS PUBLICATION, TWO-PEDAL MOTORING BY MANUMATIC, No. 444/10M/0156
7. DIETZ, R., 'A Hydrostatic Vehicle Transmission (NSU-Ebert)', *Auto. Inds.*, **9,** No. 3, 57–66 (Dec., 1964)
8. NUTT, H. and SMIRL, R. L., 'Clutches for Automatic Transmissions', *S.A.E. Trans.*, **1,** No. 4, 566 (Oct., 1947)
9. ANON, 'Fully Automatic Variable Transmission—DAF 750 Engine', *Van Doorne's Automobielfabrick publication* No. EP 4085
10. BURTON, R., 'Diaphragm Spring Clutches', *Automot. Design Engng.*, 47 (Aug., 1967)
11. HARTING, G. R., 'Design and Application of Heavy Duty Clutches', *L. R. Buckendale Lecture, S.A.E. Paper* SP239 (1962)

3

gearboxes and gearchanging

3.1 INTRODUCTION

Most vehicle transmissions include a gearbox at some point in the line of power flow. As described in Chapter 1 the variable nature of vehicle resistances due to load and gradient changes require that all of the engine's power shall be available over a range of road speeds. The low limit of this range is determined by a maximum gradient wheel adhesion, and the top limit by the rolling and air resistances of the vehicle on a level (with a light load). A gearbox has been shown to deal satisfactorily with this range of road speeds by providing a number of transmission ratios so that full power (or nearly) is available over this range (see Fig. 1.6).

Levassor is usually credited with the first acceptable gearbox for road vehicles, although Daimler was just as active with a four-speed gearbox at this time (1890–1). As seen in Fig. 3.1, the 'in line' version by Daimler is more typical of the layout which has established itself during the past 60 years. With the present tendency to economise in the space inside the vehicle, engines, gearboxes and axles are being compressed into smaller spaces. Combination of the gearbox with axle, engine, or both, shows that Levassor's layout is still extremely useful. Modern gearbox layout is remarkably similar to these early designs.

Apart from these logical layouts, having the input and output shafts 'in line', or stepped as the case of the Levassor version, the modern gearbox bears little resemblance to its earliest predecessor. Input speeds, input torques, gear tooth design, materials, bearings and other components have changed widely during these 60 years. In particular, both of these designs were of the sliding gear variety.

Gearwheels on splined shafts were slid along the shaft to engage with another gear wheel to obtain the required ratio.

In these arrangements, any misjudgment of a gearchange resulted in damage to the edges and faces of the gear teeth which deteriorated rapidly as a result of such treatment. Although the

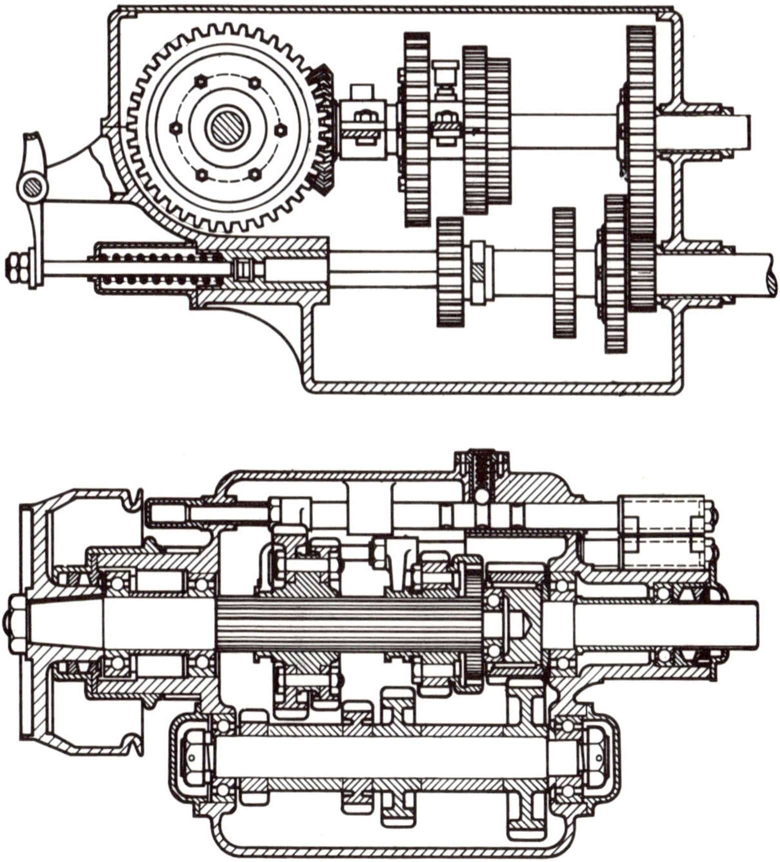

Fig. 3.1 Gearboxes from the early pioneers. Top—Levassor; bottom—Daimler

sliding gear technique is still employed occasionally for reverse gear, when an idler gear is slid into engagement, the practise is seldom used for vehicles on the move.

It has been found more suitable to keep the gears in mesh and to employ a separate sliding member to connect the required gear-wheel by means of teeth on the side of that gearwheel. The drive into the gearbox is brought via a primary gear to a gear cluster or series of pinions and gearwheels rigidly connected together on a layshaft. The gearbox mainshaft supports a number of gearwheels mounted on bearings, and in mesh with the pinions on the layshaft. When the layshaft rotates, all gearwheels in the gearbox revolve, but only one pair of gears transmits power. The selection of a gear ratio involves the connection of one gearwheel to the mainshaft by means of a sliding dog or clutch member.

This type of arrangement is referred to as a constant mesh gearbox, and the basic layout is shown in Fig. 3.2. Input to the

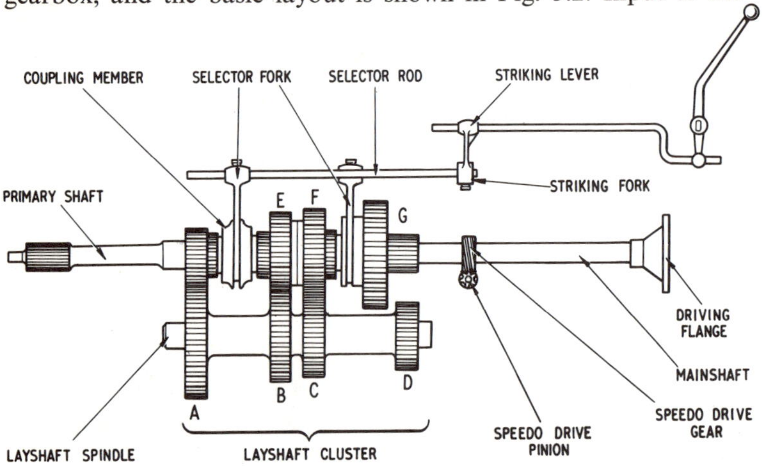

Fig. 3.2 Basic layout of a 'Constant Mesh' gearbox

gearbox comes via the primary shaft (input shaft) and drives gears *A*, *B*, *C* and *D*. Gears *E*, *F* and *G* are mounted on bearings on the mainshaft so that the mainshaft is not driven until the coupling member mounted on splines between gears *A* and *E* is moved by a selector fork to connect with the teeth on either of these gears. If the coupling member is moved towards gear *A*, then the input shaft and mainshaft are directly connected together for 'top' gear. If this member is moved towards gear *E* and engaged then the mainshaft is driven via gears *B* and *E* at a ratio dependent on the tooth numbers in these wheels. Similarly, a second coupling member between *F* and *G* makes it possible to engage the gears *CF*, and *DG*, the latter being a sliding gear engagement.

In a typical gearbox of the constant mesh type, there would be four or more pairs of gears and two or more sliding coupling members such as described above. One such coupling can be seen in Plate 3.1 which is a 4-speed commercial vehicle gearbox. The engaging teeth on the sides of the gears can be seen on each side of the coupling which itself is mounted on and able to slide on splined teeth, machined on the gearbox mainshaft. The sliding coupling is moved by a selector fork which fits in a groove in the coupling member and is attached to a selector rod mounted at the top of the gearbox.

This gearbox also shows a second selector fork which moves a

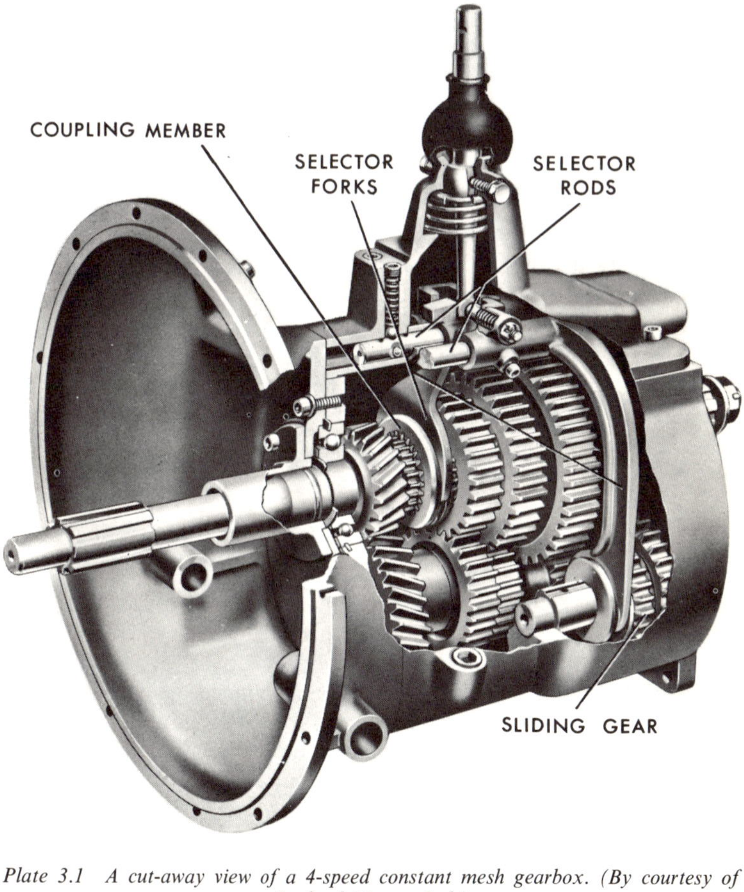

Plate 3.1 A cut-away view of a 4-speed constant mesh gearbox. (By courtesy of Leyland Motors Ltd.)

pair of sliding gears into engagement with their mating gears. This technique is still used for low and reverse gears in many gearboxes.

A particular gear ratio is selected by moving one of these selector rods forward or backward from a central (neutral) position. As seen in the illustration, the ends of the selector rods are formed so that they provide a slot to locate the 'striking fork' which the striking lever, attached to the gear lever shaft, may move. This small lever moves across the gearbox to engage with any required selector rod and in a fore and aft direction to move the

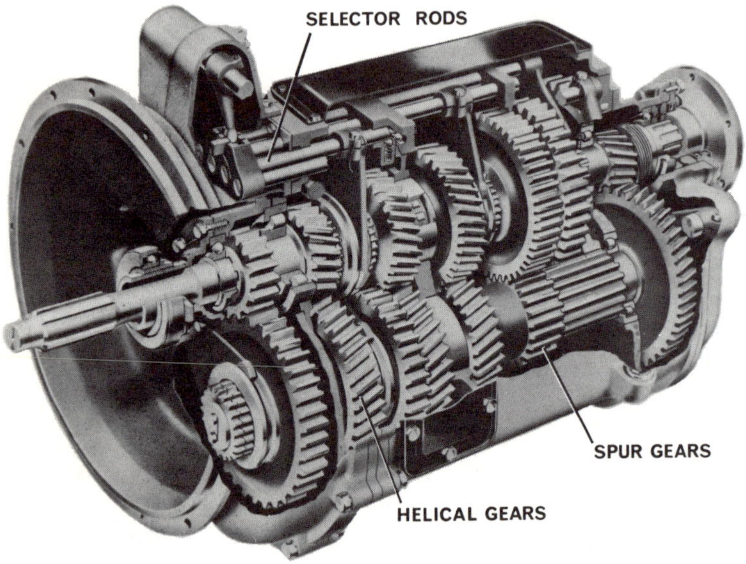

Plate 3.2 Internal details of a constant mesh gearbox showing helical and spur gears. (By courtesy of Leyland Motors Ltd.)

sliding coupling members to engage the appropriate gear ratio. The selector rods are located in any of their positions by a number of spring loaded balls which engage with grooves or depressions machined in the rods at suitable points. In a larger gearbox such as shown in Plate 3.2, the large number of sliding forks employed may require some additional sliding rods to be included to support and position the extra forks. Such additional rods are connected by linkages or other connections to the gear lever operated rods shown in this illustration at the top of the gearbox.

This illustration shows other components and features typical of gearbox manufacture. Some of the gear wheels employed for the lower gears (high reduction ratios) have straight cut gear teeth (spur gears) whereas the higher gears have gear teeth cut in a helical path across the gear face. The latter are termed helical gears and frequently employed in modern gearboxes. The helical gear has the advantage that there is no chance of intermittent tooth to tooth operation, because with the helical teeth, there are usually two teeth in engagement at any time. The helical tooth introduces side thrusts on the gear wheels which must be supported or located against this thrust but their smoother working action tends to give less risk of noisy operation.

3.2 DESIGN OF GEARS

Much of the success of the gearbox in holding its place in automobile design is due to improvements in the design and manu-

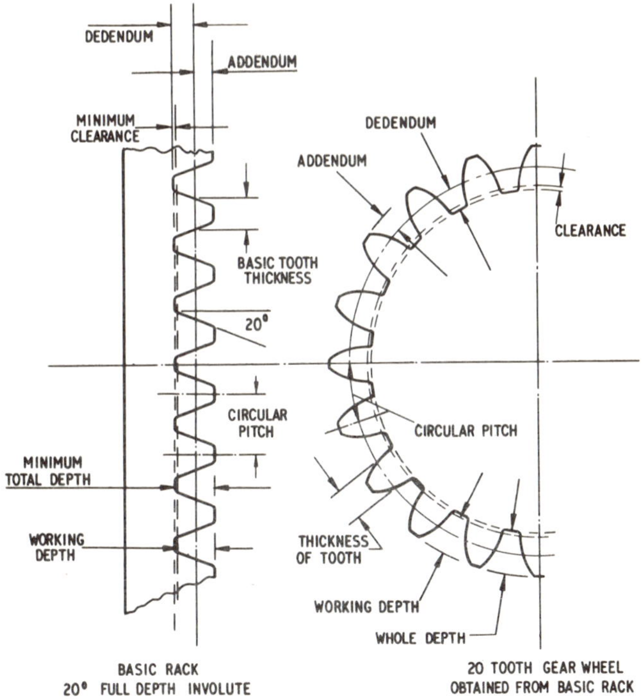

Fig. 3.3 Involute gears are based on a simple straight sided rack which can be used to manufacture gear wheels

facture of gearing. Although still conforming to the original involute tooth shape which relates to a simple straightsided tooth rack (Fig. 3.3) there have been many small changes and refinements which have enabled the gear tooth loads to rise continuously.

Slight changes such as 'correction' in which the teeth of small gear wheels are thickened by making small increases in centre distances of a gear train have produced surprisingly large increases in gear strength. Probably more important, however, have been the improvements in production methods, improved machines, more consistent heat treatment, improved inspection, etc., so that variations in production have been much reduced.

As a result, failures in gearing are probably less frequent than the bearings, shafts and castings that support them. It is not proposed to deal with the detailed design of gearing at any length but various references to literature are provided at the end of the chapter. Modern automobile gearing is, however, amongst the most highly loaded of all gearing applications. To achieve these high loads, a high manufacturing accuracy of all other supporting components as well as the gearing is necessary. Gears are made in case hardened steels so as to achieve the maximum strength and surface hardness.

Design conditions of the greatest importance are the bending fatigue strength of the gear teeth and the 'pitting failure' of the surface of gear teeth with a surface fatigue condition. Both these conditions relate to a mixture of high and low level repeated-loads on the gear teeth, but gearbox design standards are such that neither type of failure occurs with any great frequency in the current life of a gearbox.

Extremely high loads may lead to 'surface scuffing' failures, but lubrication techniques can deal with this problem which is mainly related to misuse or to unusually heavy loadings in the early life of the gearbox.

Plate 3.3 shows a part sectional view of a 6-speed gearbox revealing the arrangement of components on the mainshaft. The circular flange on the front of the gearbox is termed a bellhousing and is bolted on to the engine flywheel cover. It gets this name from its shape, which is made thus because it frequently houses a fluid coupling, or friction clutch mounted on the engine flywheel. Lubrication is usually by 'splash' obtained by gearwheels dipping into gearbox oil filled to a suitable level in the casing. Preferably, a small pump is included to ensure a feed of oil to buried shafts or bearings.

The technique of moving these sliding members into engagement with the teeth or 'dogs' on the side of the gearwheel requires just as much skill and judgment as the earlier gearboxes in which the gears themselves were moved into engagement. Efforts to reduce this skill and avoid the damage and wear, which inevitably results after a time, of the engaging parts, has led to a series of 'synchromesh' devices. Such devices all consist of a frictional cone or member which is inserted between the sliding member and the engagement teeth on the gearwheel and some synchromesh members are shown in Plate 3.3.

Fig. 3.4 is a sectional view of a 4-speed gearbox intended for a motor car and shows a typical synchromesh mechanism. Similar in layout to the previous illustrations, it will be seen that the gearbox input shaft is driven by a friction clutch assembly mounted on the

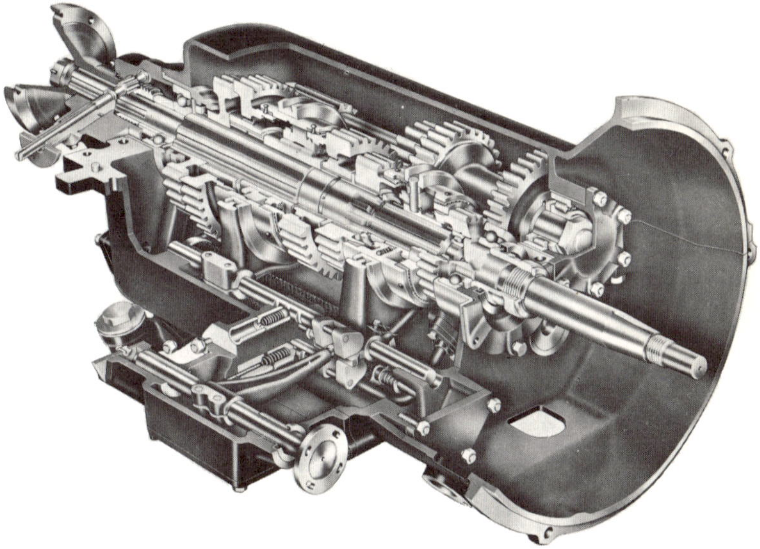

Plate 3.3 Sectional view of a 6-speed gearbox showing mainshaft details. (By courtesy of Leyland Motors Ltd.)

engine flywheel. This input shaft is supported by a large ball-bearing mounted at the front of the gearbox. This shaft is also supported by a small ball-race mounted in the end of the crank-shaft. Inside the layshaft drive gear (primary gear) is a needle-roller bearing which supports the end of the gearbox mainshaft. The primary gear drives the layshaft cluster which is a set of four gears made as one part and supported on bearings.

The layshaft gears drive mating gears mounted on the mainshaft. The three mainshaft gears are free to rotate on the mainshaft but a fourth 'reverse' gear is mounted on one of the two sliding members which are splined to the mainshaft. In this synchromesh version, the coupling member is made in two parts. The inner part is free to slide on the mainshaft splines, and supports its outer part which in turn can slide on splines on the inner part. The inner part of the coupling member contains a number of spring-loaded balls not

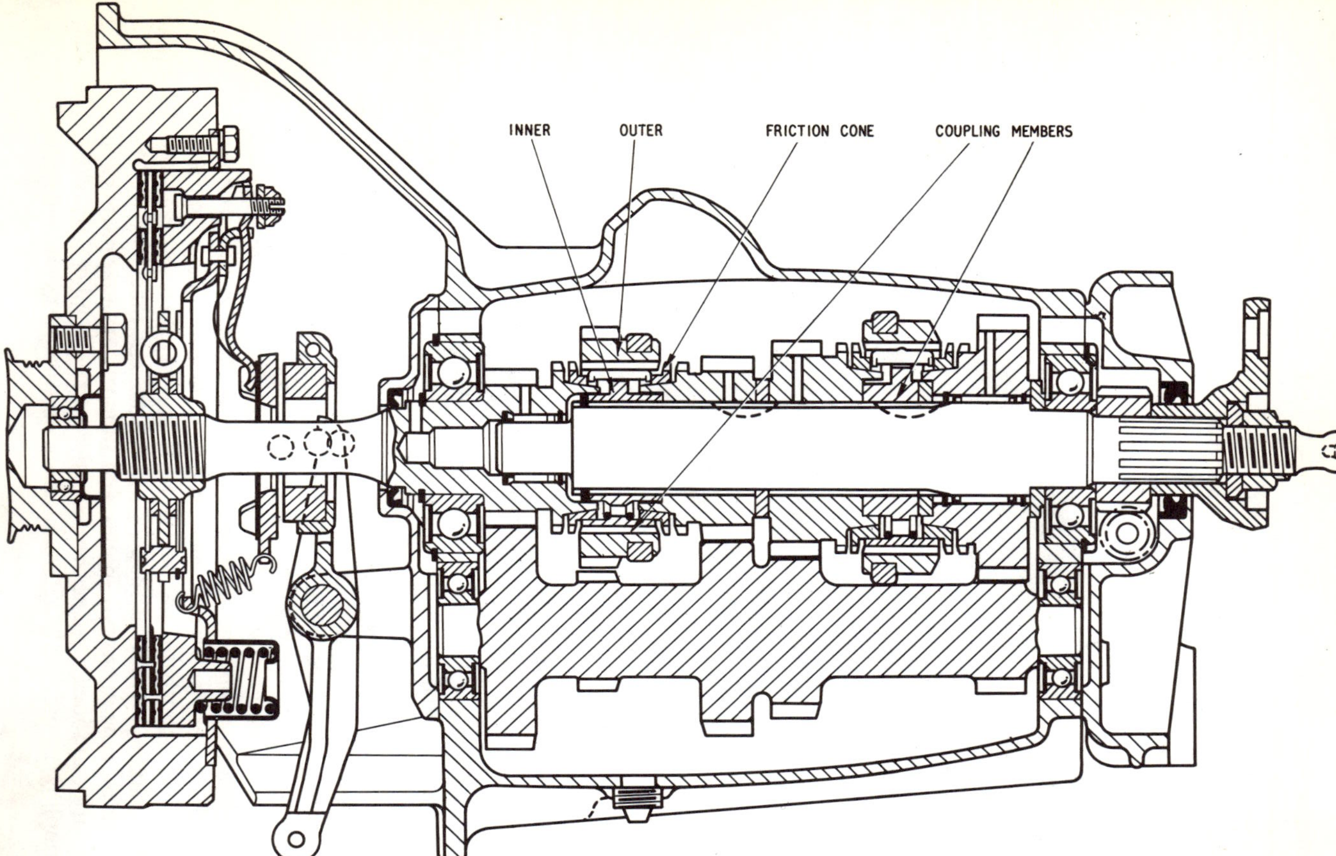

Fig. 3.4 Sectional diagram of a typical 'synchromesh' gearbox

shown here, which locate and centralise the outer member. The sides of the inner part of the coupling member are cut away to receive a cone shaped protuberance on the side of the engagement teeth of the adjacent gear. A friction cone is also mounted in the inner part of the sliding coupling and when the inner part is moved towards the gear, these friction cone faces engage with each other and synchronise the speeds of the engaging parts. The friction cones are both attached to a ring of teeth having the same dimensions of the splines on the inner/outer parts of the coupling member, so that further pressure on the selector fork depresses the spring-loaded balls and permits the outer part of the coupling member to slide across and connect with the teeth on the gear.

Movement of the selector fork to engage a required gear ratio occurs in two steps. First step consists of sliding the complete coupling member towards the gear to be engaged. This step is completed as the two friction cones engage when the inner part of the coupling is unable to move further. Further movement of the selector fork depresses the balls permitting the outer part of the coupling to engage with the teeth on the engaging gear. The spring loads, and the shape of the ball groove on the outer part of the coupling may be varied so that a higher pressure must be exerted on the selector fork and hence on the friction cones before the balls collapse.

There are a number of different types of synchromesh systems available and in use. Friction cones tend to wear in use and there is a tendency for most synchronisers to lose their effectiveness with age. Efforts to improve synchronisation has led to the baulking type of synchroniser. In this type the friction cone mounted on the coupling member is loosened or permitted a small amount of relative angular rotation relative to the coupling member. When this cone is pressed against the friction cone on the gear it is synchronising, this pressure develops a friction torque which tends to slow down or speed up the gearwheel. The same friction torque displaces the loose frictional cone, causing it to twist relative to the coupling member on which it is mounted. Baulk teeth or serrations on the outer edge of the friction cone are positioned so that these teeth prevent the outer part of coupling member to slide into engagement when the friction cone is displaced. When the gearwheel has been properly synchronised the friction torque disappears so that the friction cone moves back to its neutral position, permitting the outer part of the coupling member to slide over the teeth on the cone and to engage with the engagement teeth on the gearwheel.

In the case described and illustrated in Fig. 3.5 the 'baulking' of the movement of the outer part of the coupling member is achieved by the teeth shown on the outer edge of the friction cone. Other examples of baulking mechanisms consist of internal or external projections which prevent the movement of the coupling member. The friction cones consist of male and female cones, and there are

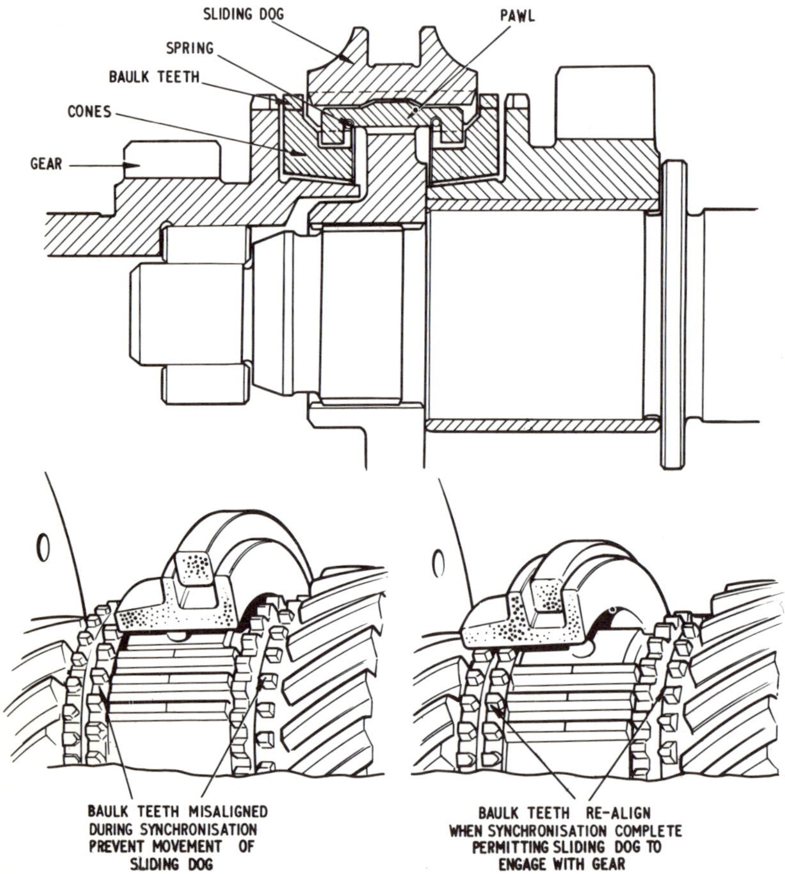

Fig. 3.5 Typical modern baulk synchromesh

alternative possibilities in which the female cone is attached to the gearwheel instead of the male cone as seen in the previous illustrations.

A further interesting synchromesh system is the servo-type such as the Porsche design (shown in Fig. 3.6). The basic elements of the coupling member are present in the servo synchromesh although

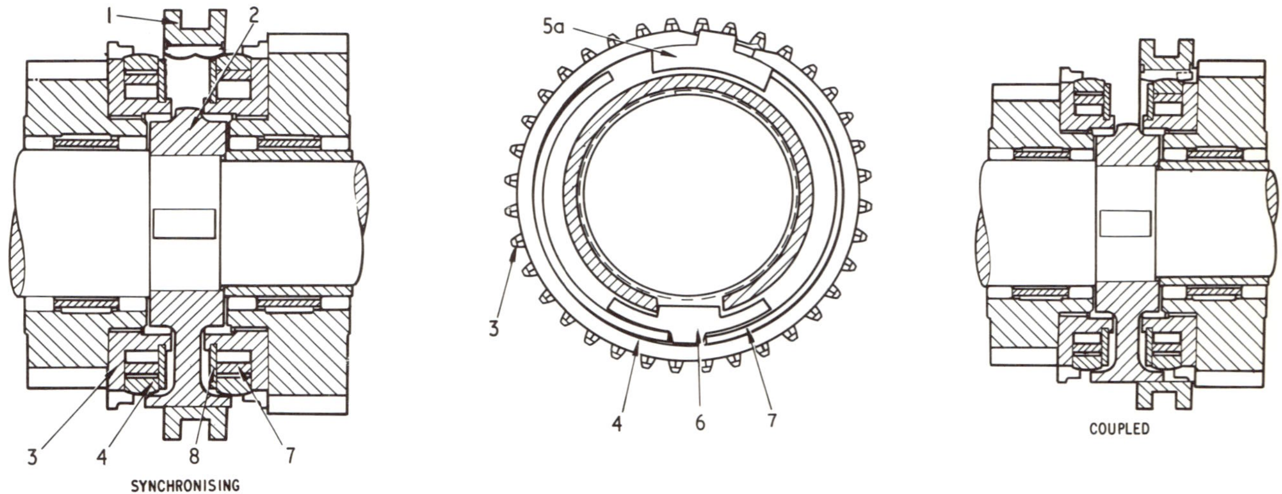

Fig. 3.6 Details of a servo-synchromesh. (Porshe)

in the Porsche design they have been reduced to a minimum width for compactness. The inner member (2) has three abutments on its outer face and the outer part (1) of the coupling member slides on these three abutments. The engaging teeth on this outer part also act as the female cone and give a further reduction in width. The friction 'cone' (4) on the gearwheel is a split ring retained by a circlip (8). This ring (4) has a chamfered outer edge to engage with the outer part of the coupling member (1). When these two parts engage and a friction torque is developed, this torque tends to unwrap the split ring which is also held outwards by an inner ring (7), causing further friction and hence further opening-up of the

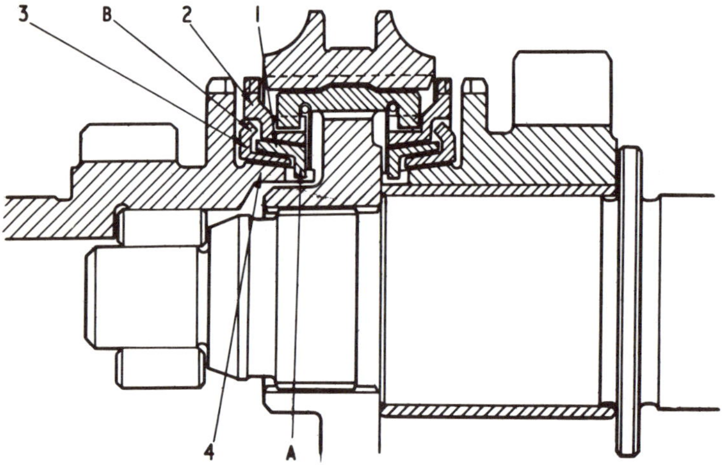

Fig. 3.7 Multi-cone synchromesh. (Smith's)

ring gap. This unwrapping effect is self generating, and is multiplied by the inner ring (7) which holds the split ring (4) outwards using the friction-torque obtained from (4) by means of member 5*A* which tends to move around the shaft towards stationary member (6). The inner ring (7) tends to 'bow' outwards and causes rapid synchronisation. The inner ring (7) collapses as soon as the gear is synchronised permitting the outer part of the coupling to move over this ring and engage with the teeth on the gearwheel.

The multi-cone synchromesh is a recent innovation which provides two additional friction cone surfaces. The illustration (Fig. 3.7) shows how two extra coned members (1) and (3), are situated between the usual male and female cones (2) and (4) respectively. The extra coned members have a number of tongues

A and *B* which engage in slots in members (1) and (3) respectively. The friction torques developed on these extra cones are thus added to that developed on numbers (2) and (4). These tongued connections also, of course, ensure that cones (1) and (3) are rotated at the same speed as the shaft and gear respectively. In this way the synchronising friction is developed on three pairs of sliding surfaces and the friction torque developed is greater as a result. Gear lever loads or synchronising time may be correspondingly reduced.

The introduction of the extra cones is skilfully done with little or no increase in size of the sliding member. The usual baulking members can be included in the multi-cone synchromesh which is complicated only to the extent of the extra cones, and offers a useful improvement to synchromesh gearboxes.

The majority of gearboxes for motor cars are still based on the synchromesh gearbox which is still subject to further refinement to improve its quietness and smoothness of operation and of its gear-changing. Difficulties tend to increase with engine speed and size, but this gearbox (see Fig. 3.8) was produced for large Jaguar engines producing nearly 300 lb ft of torque. This design uses an inertia lock type of baulk ring synchromesh on all forward gear ratios. The inset drawing shows the sliding coupling parts and both synchromesh cones. The friction cones themselves are loose-fitting and retained in the outer member of the coupling by three rectangular lugs on the friction cone, these lugs fitting in a slot in the coupling. The slot being wider than the lug, friction cones tend to move within the slot when they are loaded into engagement during synchronisation. This movement in the direction of the friction torque provides a baulk because a number of serrations, or teeth, on the outside of the friction cone 'baulk' the movement of the outer part of the coupling. When synchronisation is complete, pressure on the baulk teeth due to the outer part of the coupling member moves the friction cone to the central position permitting the coupling to slide over the friction cone and engage with teeth on the synchronised gear.

This gearbox design makes extensive use of needle-roller bearings to support loaded gearwheels and shafts, etc. Also, the in-built internal gear pump at the rear of the gearbox is less commonly used on motor cars, but when incorporated as in this illustration allows oil to be fed under pressure to ensure adequate lubrication to all necessary parts. The gearbox casing is made from cast iron to ensure rigidity and sound deadening and is typical of most other designs although in some manufacturers there is a tendency to introduce light alloy casings for gearboxes in order to save weight.

A precise gear lever movement is usually very popular with drivers and improvements are being made to the gear lever/selector fork operation. In the Vauxhall Viva, for instance, selector forks are arranged to slide on a single rod, and a rotary selector bobbin on the gear lever shaft provides a means of obtaining very precise positioning of the gear lever, with some reduction in manufacturing cost.

It will be clear from the previous paragraphs that passenger car applications are mainly satisfied by a four-speed gearbox. The engine has an adequate reserve of power to give the necessary

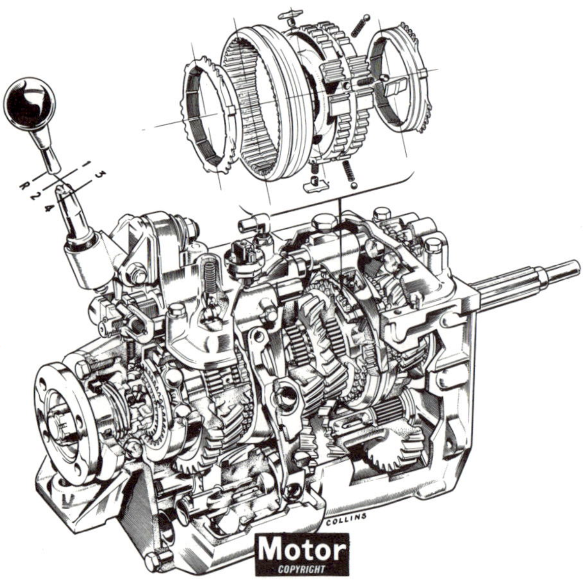

Fig. 3.8 Sectional illustration of a synchromesh for a high powered car. (Jaguar)

acceleration or hill climbing ability for most road conditions. Axle or gearbox ratios are usually chosen so that normal cruising is carried out at comparatively low engine torque, thus leaving extra power to be available when the accelerator pedal is depressed. Emphasis in motor car gearboxes is mainly given to simpler and smoother gearchanging, especially by using improved synchronisers to obtain light and precise gear lever movements.

Gearbox synchronisers have an apparently simple duty, since with the clutch disengaged, the friction cones are required to accelerate or decelerate only the friction plate in the clutch, together with the input shaft and the layshaft gears. In practise, however, it is found that these parts have sufficient rotary inertia to impose a difficult duty on the friction cones. The energy lost by the synchronised parts or the energy supplied to them is transferred via the friction cones which receive the same amount of energy as heat.

3.3. SYNCHRONISER THEORY

The process of gear synchronisation involves a change in the energy contained in the spinning clutch disc or plates. Although these parts are made as light as possible for this reason, these still have some rotary inertia and the clutch plate, together with the shafts and any gear wheels rotating with it, must be considered as a small flywheel. When a gear change is made, the clutch plate is accelerated to a high speed (for a down-change) or is slowed up to change to a higher gear. This process involves a change in energy contained in the rotating parts. Energy must be supplied for acceleration, or alternatively, energy will be given up when these parts slow down. In either case the transfer of energy occurs through the synchroniser device.

One cone (or part) of each synchroniser is attached to the output shaft and is connected directly to the driving wheels. This cone may thus be considered as being attached to a very large flywheel equal to the total inertia of the vehicle. The second cone (or element) of each synchroniser is attached via gearing to the clutch plate, and the ratio of gearing depends on which speed in the gearbox is about to be connected. In normal gear changing, when the next higher or lower gear ratio is selected, the speed of the clutch plate is likely to be altered by 20%–40% from its original speed of rotation.

The process of synchronisation can be likened to the action of a simple cone clutch which is used to connect two flywheels together. One larger one (represents the vehicle) will be connected to a much smaller flywheel (representing the clutch plate and its associated gearing) which may be rotating faster or slower than the large flywheel by 20–40%. As the elements of the synchroniser are brought into engagement, the speeds of the two flywheels will be altered until, at engagement, they both rotate at exactly the same speed. Heat will be generated in the synchroniser cones during this process so that by considering the energy of the system:

$$I_1\omega_1^2 + I_o\omega_o^2 = (I_1 + I_o)\omega_s^2 + H \tag{3.1}$$

where I_1 is the rotary inertia of the clutch plate etc. (small flywheel), I_o is the rotary inertia of the vehicle (large flywheel), ω_1 is the initial speed of rotation of the small flywheel, ω_o is the initial speed of rotation of the large flywheel, ω_s is the final speed of rotation of both flywheels, H is the heat loss during engagement.

In practise there is only a very small change in the speed of the vehicle (large flywheel) during engagement and also it can be shown* that the slipping clutch operation has an efficiency of about 50% for the transfer of energy.

The change in energy in the clutch plate is therefore:

$$\text{Energy loss} = I_1(\omega_1^2 - \omega_s^2)$$

When ω_1 is greater than ω_s then the clutch plate gives up energy through the synchroniser to the vehicle which increases speed very slightly (up-change). The vehicle receives a small increase in energy equal to half the energy lost by the clutch plate. The up-change 'jerk' and the heat loss are both equal to $\frac{1}{2}I_1(\omega_1^2 - \omega_s^2)$.

In the case of a down-change the clutch plate has to be accelerated to a higher speed because ω_s is greater than ω_1. The clutch plate actually receives energy from the vehicle and since the synchroniser works at only 50% efficiency the vehicle supplies twice as much energy as the clutch plate receives.

In a down-change the vehicle loses $2I_1(\omega_1^2 - \omega_s^2)$ and the heat loss is one half of this energy.

Therefore, considering a particular roadspeed, a change to a lower gear generates twice as much heat as the equivalent up-change. Also, the vehicle loss of energy is 4 times that received during the up-change.

Putting these losses into practical units, the rotary inertia I_1, of a clutch plate and connecting parts is not likely to be more than 0·005–0·01 lb ft sec^2 units, but the speed change may be from 4,000 to 3,000 rev/min (perhaps greater), hence $\omega_1 = 400$, $\omega_s = 300$ approx. so that the clutch energy change would be 350–700 ft lb of energy. Taking a mean value of 550 ft lb then the various energy changes are shown in Table 3.1.

Table 3.1

	Up-change	*Down-change*
Clutch Plate	loses 550 ft lb	gains 550 ft lb
Vehicle	gains 275 ft lb	loses 1,100 ft lb
Heat Loss	275 ft lb	550 ft lb

*see, for instance, bibliography reference number 10.

The most serious loss (1,100 ft lb) is equal to a 2 hp loss of power for a period of 1 sec and is noticeable as a jerk, but the engine can supply the necessary clutch energy, and the vehicle is usually slowing anyway, but it is easy to see how these energy changes become more important in automatic gear changes where the rotary inertia of the engine and flywheel may be 20–100 times the inertia of the clutch plate.

Also, it will be noted that the synchroniser cones absorb 550 ft lb of energy. This is equal to transmitting 4 hp if the synchronisation is completed in 0·25 sec and is a fairly high rate of working. The rise in temperature of the cone parts can be estimated from their weight and unless heat is removed quickly, repeated gear changes would lead to very high temperatures especially on the working faces of the synchroniser cones. An increase in the size of cones would give an immediate improvement in many respects, but introduces a cost penalty. The problems of synchronisation are mainly due to deterioration with use. The heat and wear problems clearly increase with the size of the vehicle which requires large plates, gears and shafts, but also increase more rapidly as engine speeds tend to increase.

The requirement for smooth and simple gearchanging leads naturally therefore to the use of automatic gearboxes and transmissions. In these designs, the driver is not required to change gear, except in unusual conditions, but if he does wish to move the gear selector all synchronising is carried out by friction clutches or brakes which have sufficient capacity to smooth and synchronise all rotary parts including the engine. These aspects of full power synchronisation are considered in more detail in a later chapter. It is important to appreciate, however, that efforts to improve and reduce synchronisation time involve the same problems as in automatic gearchanging.

Although friction cone synchronisers have been included in commercial vehicle gearboxes, they are by no means universal. The rotary inertia of the synchronising parts is obviously greater than their equivalents in the passenger car, but it is also possible to use larger friction cones in the commercial vehicle gearbox. Even so, it is not always usual to synchronise all speeds and commercial vehicles mainly rely on the driver's skill with a constant mesh gearbox.

Commercial vehicles require more gear ratios than passenger cars because their diesel engines have a smaller speed range than the petrol equivalent and also because they require to use full engine power more frequently than the average motor car. Hence it is by no means unusual to find 8-, 10- or 12-speed transmissions in use on heavy commercial vehicles. Any vehicle which requires to have maximum engine power available over a wide range of road speeds needs a larger number of gear ratios, so that high performance sports, and racing vehicles, also show the same trend to increase the number of available gear ratios.

Clearly, as the number of gear ratios in the gearbox increases, the size, weight and cost of transmission will all likewise increase. Such increases are, of course, a disadvantage in all forms of road vehicles, but in the case of racing cars some very ingenious methods have been adopted for reducing the size and weight of the gearbox. As seen in Fig. 3.9 a 6-speed gearbox and axle with integral oil tank are designed into a very small space.

Input from the racing engine is taken right through the gearbox at bottom by means of a long shaft which extends to an input reduction gear train mounted at the right hand side of the gearbox. This reduction train reduces the input speeds from over 10,000 rev/min to around 6,000 rev/min for the gearbox drive shaft which runs outside and concentric with the long input shaft. The six forward speeds and a reverse gear are obtained by seven pairs of gears together with a reverse idler gear, all of which are contained within a space only just over 6 inches in length.

The gear change is of the simple dog-engagement type with teeth machined on the side of the gearwheels. Any of the required ratios are engaged by means of a coupling member which slides on a splined shaft to engage one gear of the required pair. This technique is similar to that in some of the earliest designs, but in this case, it will be noted that the coupling member is in itself a gearwheel. Thus, on the right hand side of the gearbox compartment a coupling member mounted on the gearbox output and axle-pinion shaft slides forward or backward to engage the first and second speeds in the gearbox. This coupling member carries gear teeth on its outer periphery and when the coupling member is in its central (neutral) position a reverse idler may be moved into mesh with the coupling teeth and also with the teeth on a narrow gear mounted on the lower gearbox input shaft. This coupling member enables reverse gear to be obtained in a simple way. It will be noted that instead of the usual selector fork, the coupling member is moved backwards or forwards by means of a sliding 'cap'-member which fits over the gear teeth on the coupling member and exerts pressure on the side of the gear teeth to move this member.

The technique of geared coupling members is also used on the left hand side of the gearbox compartment which includes two further geared coupling members, one on each of two gearbox shafts. The coupling on the output shaft is employed to engage

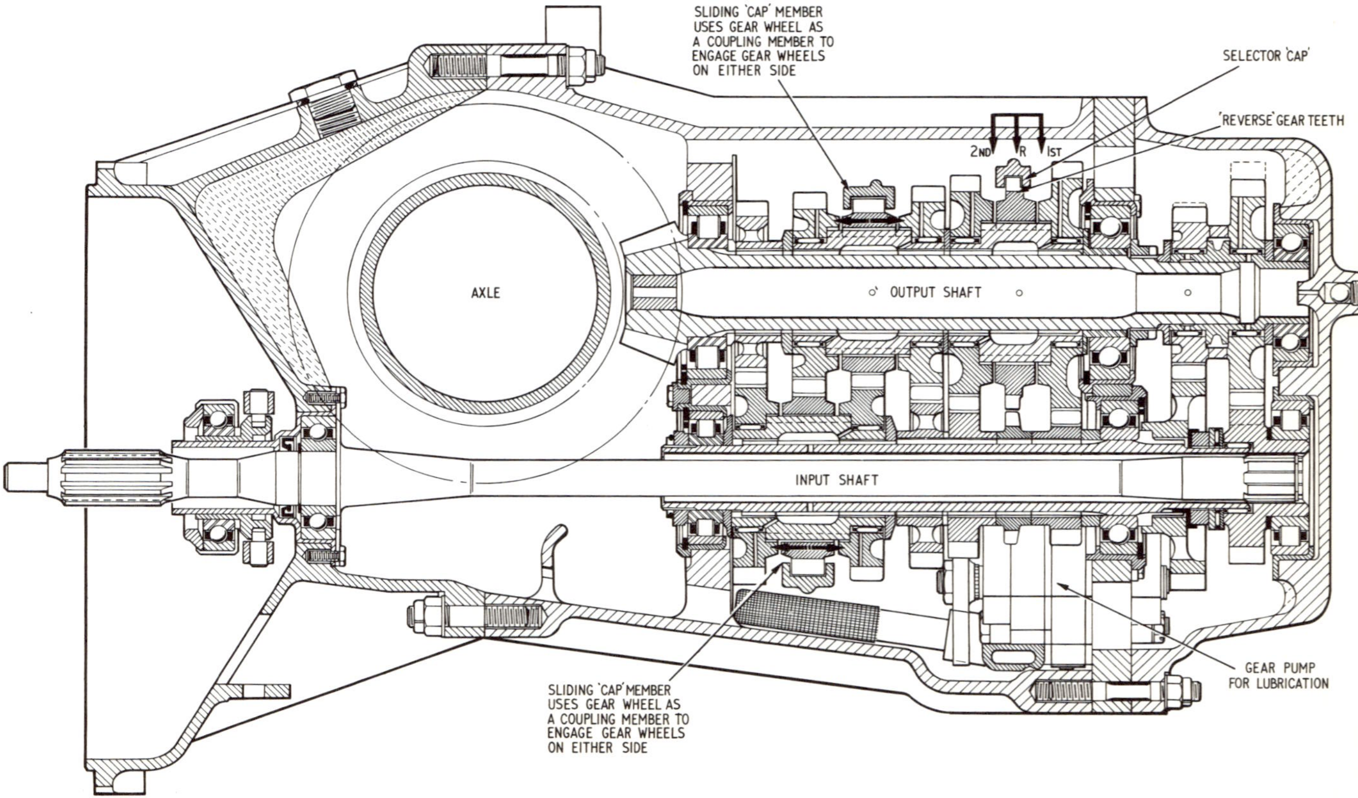

Fig. 3.9 A compact 6-speed gearbox and axle for a racing car. (Rubery Owen—B.R.M.)

third and fourth speeds whose gear wheels are arranged as the third and fifth pairs from the left of the gearbox bulkhead. This coupling is in fact a sliding gearwheel and when positioned in the centre (neutral) of its travel acts as the driven gear for fifth speed (fourth pair from right).

In a similar way the lower coupling on the gearbox input shaft is employed for changes between fifth and sixth speeds, but being manufactured as a sliding gear, acts as the driving gear for fourth speed when positioned in the centre of its travel. In all cases the gearwheel-coupling member is moved by an internally grooved cap which fits over a segment of the rotating gear. As a result an extremely compact gearbox is achieved, together with a corresponding reduction in weight. Another very compact racing car 'transaxle' is shown in Fig. 3.10.

When a greater number of speeds or gear ratios than six is considered, greater saving can be achieved by dividing the gearbox into two parts which are used in a series combination. Thus, for instance, the six pairs of gears in a 6-speed gearbox could be re-arranged into two 3-speed gearboxes. If the gear ratios are chosen suitably it would be possible to operate the vehicle as a 9-speed gearbox because all three of the gear ratios in one gearbox could be used in combination with any of the three gear ratios in the second gearbox. Many heavy commercial vehicles incorporate two gearboxes in this way. In some cases the second gearbox is called a splitter gearbox and is intended to offer a number of small steps between each ratio in the main gearbox. In other cases an auxiliary gearbox extends the range of ratios in the main gearbox by introducing one or more fairly large ratio steps.

Having decided to obtain a large number of gear ratios by means of two gearboxes, it is possible to put forward a large variety of alternative solutions. In many cases auxiliary gearboxes or splitter-boxes are designed with a particular main gearbox in mind. The gear ratios in the second gearbox are chosen, therefore, to suit the gear ratios and steps which are already available. In some cases there will be overlapping so that the same ratios are obtainable by more than one combination, and in other cases certain gear ratio combinations are not of great value. Gear lever operation, too, must be considered since the driver may have to move two levers at some time to achieve the required gear reduction ratio.

One example of a 10-speed gearbox which was designed from the start in the form of a 5×2 combination is shown in Plate 3.4. A 5-speed constant mesh gearbox is combined with a 2-speed range-change synchromesh gearbox. The sequence of operation is to proceed from speeds 1 to 5 in the main gearbox, then to change

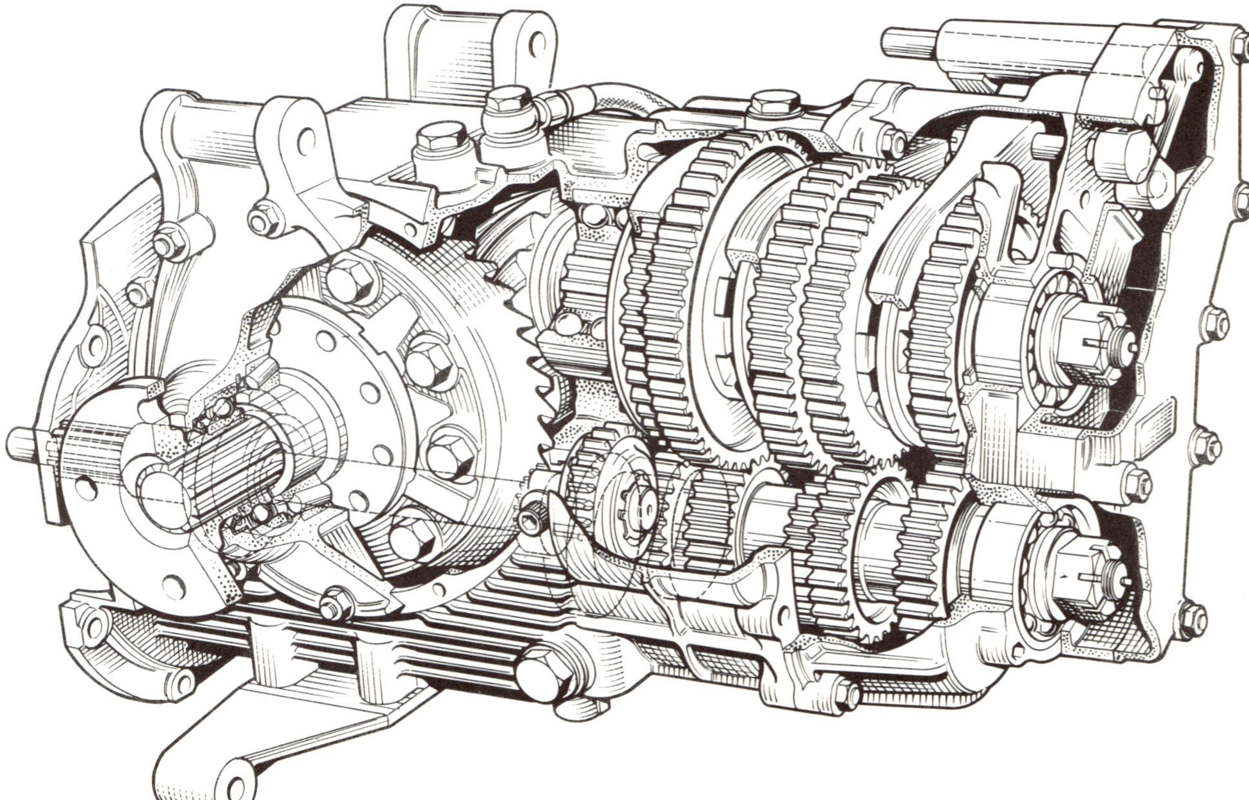

Fig. 3.10 A 5-speed racing car gearbox (Zahnradfabrik Friedrichschafen AG)

from low range to high range in the auxiliary gearbox and then continue gearchanges 6–10 which are made by retracing the first to fifth speed positions in the main gearbox. In this way it is possible to span a 10 to 1 range of ratios with a number of approximately equal spacing. In a typical gearbox as shown in this figure the actual gear ratios are as shown in Table 3.2.

Table 3.2

Speed	1	2	3	4	5	6	7	8	9	10
Gear Lever Position	1	2	3	4	5	1	2	3	4	5
Range Lever Position	Low (3·54 :1 Redn)					High (Direct)				
Gear Reduction	9·65	7·43	5·80	4·48	3·54	2·73	2·10	1·64	1·27	1·00
Percentage Reduction	30	28	29	27	30	30	28	29	27	—

It will be noted this 10-speed gearbox has been designed to obtain practically equal steps of 27–30% reduction with each downchange. Gear designers will know that this is a difficult operation because of the limited choice of gear tooth numbers which are normally available.

To assist the driver, a ratio change in both gearboxes is required only between fifth and sixth speeds and the range-change lever is a small knob which operates a pneumatic cylinder to change from high to low range, or vice versa. The range-change gearbox includes some multi-disc type synchronisers to avoid any 'crashing' of the engaging dogs as a result of the power shift. The range-change knob shown on the side of the gear lever is normally operated when the main lever is in the neutral position and being moved from one side of the gear lever 'gate'. The range-change is quickly effected by the pneumatic cylinder, and this introduces a minimum delay into the gearchange between fifth and sixth speeds.

Intended for a variety of transmission duties, it will be noted that the layshaft contains an additional gear wheel (second from left) which drives a gear on a power-take-off shaft (not shown). This power-take-off shaft can be used to drive any auxiliary mechanism and can be used when the vehicle is stationary or moving. The sliding gear at the right hand end of the layshaft is for obtaining reverse gear which is available with either the high or low ranges.

Having provided the driver of a heavy road vehicle with a 10-speed gearbox, it must be remembered that a lot more gearchanging will be required. The driver will be able to obtain near-maximum power at any road speed since he need never be more than 27% from the maximum engine speed. To obtain best operation, the driver will not only need to make nearly twice as many gearchanges as for a 5-speed gearbox, but will require a skilled judgment of the operation of this very close ratio gearbox. In approaching a hill, for instance, it would be advantageous if he changed down quickly from tenth to eighth, or even seventh gear because the time lost in progressively changing through each ratio would lose too much of the vehicle's speed and impetus up the gradient.

A multi-ratio gearbox for heavy vehicles calls, therefore, for

Plate 3.4 10-speed 'range-change' gearbox. (By courtesy of Eaton–E.N.V.–Fuller)

additional work by the driver and for additional skill not only in judging what ratios to select for a given load and road, but also for skill in making gearchanges to a new ratio one, two, or even three steps above or below his present gear selection.

There are many engineers who consider that a multi-ratio gearbox having more than six available speeds, must inevitably be of automatic operation. Unless some automatic device is available for continually selecting and engaging the correct gear ratio, the extra gear ratios will call for too much of the driver's attention to ensure that optimum operation of the vehicle is achieved.

Thus, as in the case of the passenger car, we find that progress is pushing gearboxes towards automatic operation. Whereas in cars the desire for automatic transmission arises from smoothness and ease of operation, in heavy commercial vehicles the need for best performance from the engine leads to an increased number of gear ratios, which eventually become too numerous for manual operation and hence call for an automatic device.

BIBLIOGRAPHY

1. PHILLIPS, R. E., 'Variable Speed Gears for Motor Road Vehicles', *Proc. Inst. Mech. Engrs.* (London), Automob. Div., 833 (1916)
2. SHORT, L. J., 'Transmission Gear Developments', *Proc. Inst. Automob. Engrs.*, 337 (1937/8)
3. STOTT, T. C., 'Problems in the Design of an Economical Gearbox', *Proc. Inst. Mech. Engrs.* (London), Automob. Div., 11 (1953/4)
4. ERNST-METZMAIER, R., 'Synchromesh Systems', *J.Soc.Ing.Auto.*, **28,** No. 11, 519
5. MERRITT, H. E., *Gears,* Pitman (Sir Isaac) & Sons, Ltd. (1954)
6. SOKOLOWSKI, T. L., 'Gearbox Design Loads', *Automob. Engr.*, **55,** No. 13, 527 (Dec., 1965)
7. SCHILKE, W. E., 'The Reliability of Transmission Gears', *S.A.E. Paper* 67075, 13 ff (Sept., 1967)
8. WILSON, A. G., 'Transmission Developments for Public Service and Heavy Goods Vehicles', *Proc. Inst. Mech. Engrs. (London) Automob. Div.*, 93 (1956/7)
9. ANON, 'David Brown, 10-speed Gearbox', *Automob. Engr.*, **57,** No. 10, 375 (Sept., 1967)
10. GILES, J. G., 'Problems of Making a Gearchange Under Power', Cranfield symposium 1963. *Advanced school of Automobile Engineering paper.*
11. ANON, 'Vauxhall Viva Transmission', *Automob. Engr.*, **57,** No. 1, 16–18 (Jan., 1967)

4

automatic gearboxes and control

The trend towards an automatically operated gearbox or transmission comes, as was seen in the previous chapter, from two separate and distinct needs. In the case of passenger cars, designers are seeking for smoothness of operation and ease of control, whereas the commercial vehicle requires more and more gear ratios and a minimum interruption of power during gear changes. It will be seen later that the two differing requirements lead to a common concept, and in both forms of transport there is a tendency to adopt some form of fluid drive. Epicyclic and planetary gearing is nearly universal for automatic transmissions and wide use of friction clutches and band brakes is made to introduce different gear ratios. Moreover, having provided transmissions which can change ratio quickly and smoothly, it becomes desirable to include an automatic control to decide when a ratio-change is needed, and to execute that change. On top of this, there is the need to permit a driver to override the automatic control when he wishes.

Thus although automobile gearboxes have improved steadily over the years with general evolution of road vehicles, the transition to automatic operation has caused more of a revolution than evolution because there is very little resemblance between a modern automatic and the earlier gearboxes. Before describing complete transmissions it will be necessary therefore to consider and discuss some of the interesting new components which are commonly seen in the new designs.

4.1 FLUID DRIVES

Automatic transmissions almost invariably incorporate some form of fluid drive chiefly because of their excellent automatic operation for the starting and stopping of the vehicle as discussed in Chapter 2. In particular, a three-element fluid converter has become a common feature mainly as a result of experience from the U.S.A. These converter units have developed from the simple fluid coupling when by inserting a further reaction member in the path of the circulating fluid it is possible to obtain torque multiplication on the output member (Converter-coupling).

Although early developments attempted to replace a vehicle's gearbox completely by means of the fluid converter it is currently more usual to use a converter coupling version which combines the properties of a fluid converter with those of a fluid coupling. As shown in a previous illustration (Plate 2.1) the converter-coupling has three rotary elements known as the pump or impeller, turbine or output, and a reaction member. Considering the sectional illustration shown in Fig. 4.1, it will be noted that the fluid rotates in a circular path in the section plane in addition to the rotation bodily around the axis of the flywheel.

The pump is driven by the engine and produces the circulation of fluid with which the converter casing is filled (frequently under pressure). The pump draws in fluid at the small diameter at its centre and due to its rotation ejects the fluid at high velocity from exit vanes or blades situated at the maximum diameter around its periphery. Although these exit vanes are approximately radial in direction (i.e. parallel to engine axis) the rotational velocity of the impeller is approximately equal to the circulation velocity. The fluid leaving the impeller is directed at about 45° and strikes the vanes of the turbine so as to drive the turbine in the same direction as the impeller.

The turbine member develops output torque from the large quantity of fluid which impinges on its vanes and drives the turbine members in the same direction as the pump. By changing the angles of the turbine vanes at entrance and exit, different characteristics can be obtained but in general the turbine is designed to have high coupling efficiency and to obtain about twice times pump torque when the turbine is stationary.

A reaction member is employed to collect the fluid leaving the turbine member and to redirect this fluid into the pump for re-circulation. In re-directing this fluid a reaction torque is developed on the reaction member which must be sustained against rotation in the opposite direction to that of the pump and turbine. This

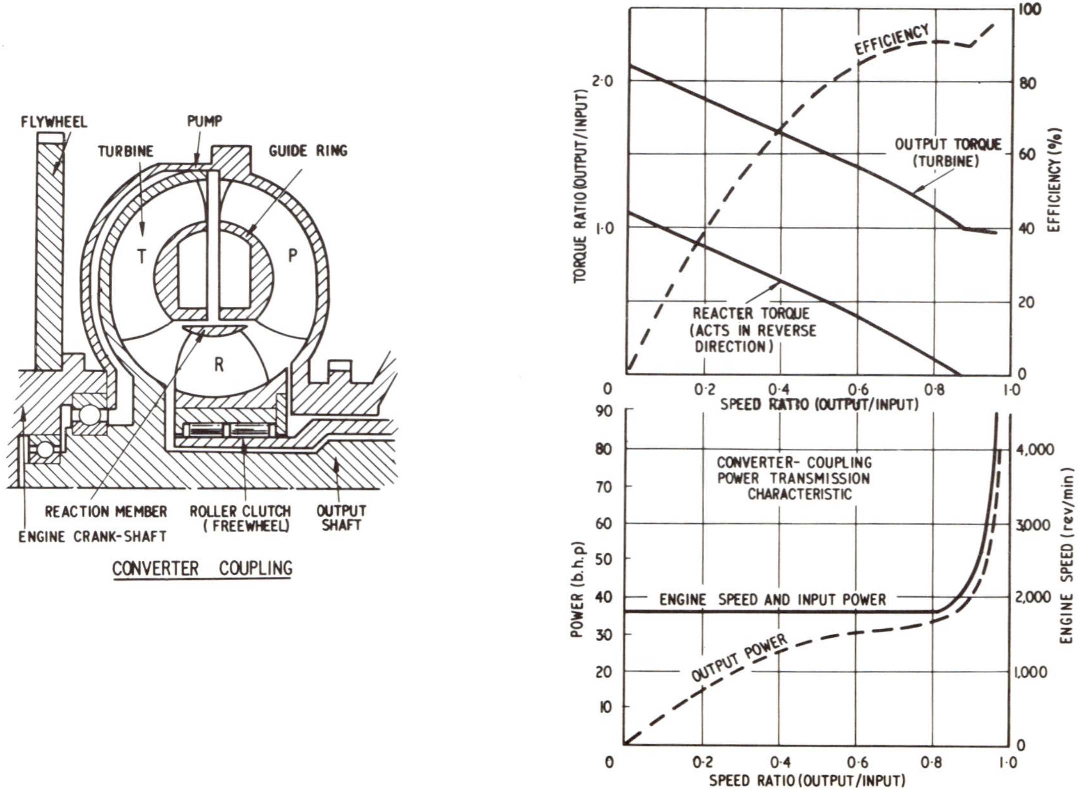

Fig. 4.1 Typical Hydrokinetic converter–coupling performance curves

reaction torque is similar to the reaction on a normal gearbox which develops a torque equal to the difference between input and output torques. In a fluid drive the reaction member is usually mounted on a freewheel device or roller clutch. This one-way clutch prevents the reverse rotation of the reaction member but permits their member to rotate forwards when the turbine speed approaches that of the pump. This forward rotation occurs when the pump and turbine torques are equal (1 :1 ratio). The freewheeling reaction member offers less resistance to the circulating fluid, and produces an improvement in coupling efficiency.

The curves in Fig. 4.1 show typical operation of modern converter couplings. It will be noted that the maximum output (turbine) torque occurs when the turbine is held stationary and this torque falls progressively as the turbine speed approaches that of the input (pump) member. When the turbine torque falls to the value of input torque, the reaction member torque is zero and this member commences to freewheel as the transmission ratio continues towards unity. The peak efficiency point of the converter operation is known as the design point and the vanes and blades in the converter are selected for this point. To obtain more output torque multiplication the design point must be moved to the lower transmission ratios, but for best efficiency in the coupling range, design points are usually near 0·75–0·85 speed ratio. The reverse torque on the reaction member decreases as the turbine speed increases and reaches zero at the coupling point.

The rate of fluid circulation is much affected by turbine speed and the torque required to drive the pump is highest when the turbine is stopped. In most converter designs the engine speed is dragged down to its maximum torque speed when the turbine is stationary, and this is known as the engine stall speed. As the turbine speed increases the pump torque stays constant and only at the higher speed ratios will the engine speed increase, as shown on the graph. It will be noted from this graph that the converter is chosen so that less than half of the engine's power can be developed at maximum torque multiplication, and the full engine power can only be transmitted when the fluid drive reaches the coupling range. Development in U.S.A. of these converter couplings has resulted in designs which limit the amount of power which can be transmitted for converter operation and have a high coupling efficiency. Both these aspects of performance result in reduced fuel consumption, but when the smaller-engined European vehicle is considered, the limitation of engine power during the torque conversion stage would give a very poor hillclimbing and acceleration performance.

European vehicles mainly require to have the full engine power available more frequently than the larger American cars and hence require torque converters which can give torque multiplication right through the engine speed range. The same requirement exists for commercial vehicles which also make frequent use of full engine power. Unfortunately, hydrokinetic torque converters which can provide their maximum torque multiplication at full engine power give increased fuel consumption and the general trend is to accept some power limitation of the converter and to employ an auxiliary gearbox giving a number of extra gear ratios, so that for most of the operating time, the fluid drive is acting as a coupling.

Many different attempts have been made to make better use of the converter range of these hydrokinetic fluid drives and one of these was to insert a gear ratio between the engine and fluid drive. This had the effect of altering the input torque to the converter and by 'changing down' and applying a higher torque, the fluid drive was influenced towards the converter range. This gear ratio also increased engine speed so that there was a substantial increase in the transmitted power for conversion.

The Voith Diwabus transmission attacks the problem in another way which employs a gear train inserted between the engine and fluid drive (see Fig. 4.2). A differential gear is connected so that the engine torque is divided approximately equally between the output shaft and the fluid drive. Some of the engine's power is transmitted directly to output without loss by mechanical means but when the vehicle is starting the output shaft is rotating slowly so that the majority of engine power passes through the fluid drive. During this starting procedure the fluid drive is accepting about half engine torque but is driven by the differential gear at nearly twice engine speed.

The differential gear—which gives the name to the transmission—represents another way of emphasising the converter operation for starting and acceleration, whilst retaining the coupling performance for normal cruising. As seen in the graph, the input speed and power to the fluid drive decreases as the output shaft accelerates. The efficiency of transmission increases also by virtue of the increased mechanical power and more efficient operation of the fluid drive.

Since converter operation is employed at about half engine torque the maximum output torque is somewhat limited and it is necessary to add further gear trains which add to the cost and complication of the transmission but it has seen excellent service in German buses and coaches.

Thus, although the fluid converters were originally intended to

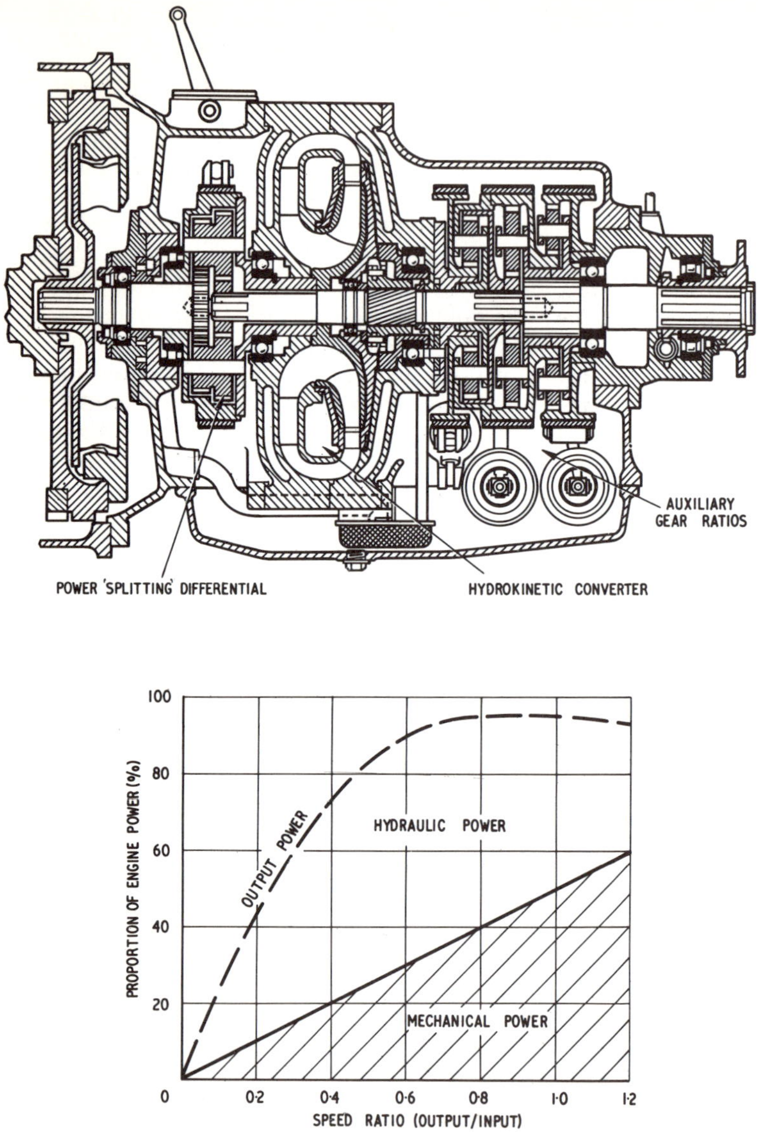

Fig. 4.2 The Voith Diwabus Transmission employs a differential gear to share the engine power between a fluid converter and gearing

replace the automobile gearbox, they have finally been employed as an improved form of starting device or clutch giving limited torque multiplication and helping to smooth any gearchange jerks in the automatic gearbox.

4.2 MULTIPLATE CLUTCHES

Automatic transmissions do not disconnect the vehicle's engine when a gearchange is made and the problems of synchronising the gears for the new ratios are much greater because of the engine, flywheel and other rotary inertias attached to these gears, and because of the torque being delivered by the engine which continues to work during the gearchange. It is clearly not possible to employ a manually operated friction cone to synchronise the engine for an automatic gearchange and instead, a more powerful type of friction clutch is needed. Clutches employed for this purpose are usually multiplate designs. The torque developed by a friction clutch depends mainly on the applied load and the coefficient of friction of the working surface. Although the use of a number of thin friction plates brings about a reduction in the diameter of the clutch assembly, the multiplate designs are chosen because of a heat rating of friction materials. When clutches slip under torque, this operation being a necessary part of the gearchange function, considerable heat is generated on the contact surfaces. Clutch plates can be rated on their heat dissipation properties and the best materials can be used up to 5 bhp/in^2 of surface.

By choosing a clutch consisting of a number of thin plates connected alternately to input and output shafts it becomes possible to obtain a very large area of working surface in a comparatively small space. Multiplate clutches usually work in oil and their

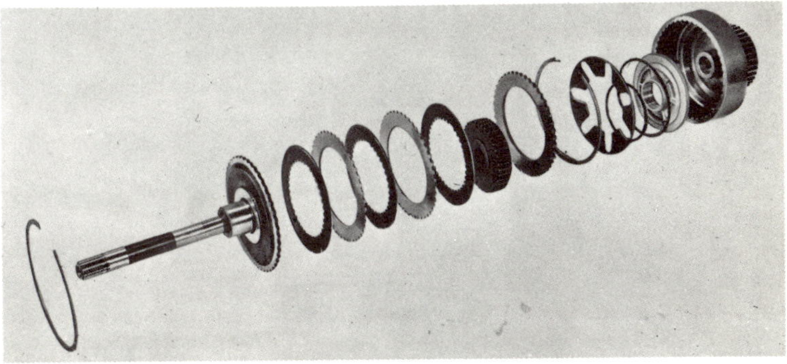

Plate 4.1 Components in a typical multiplate clutch

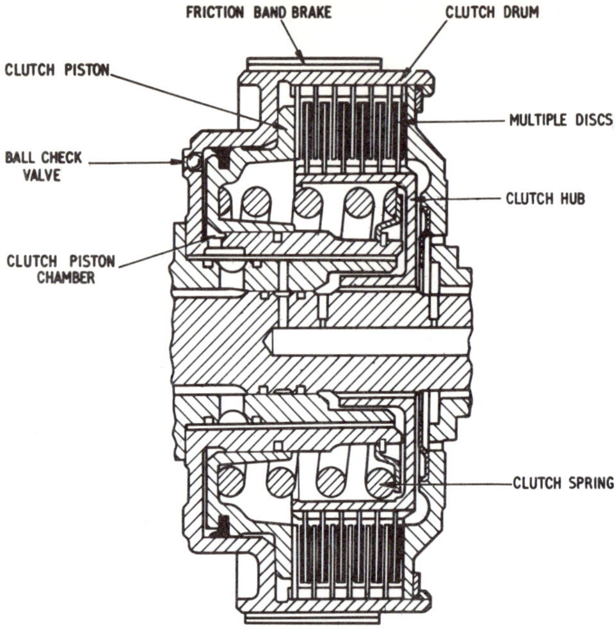

Fig. 4.3 A typical multiplate clutch assembly

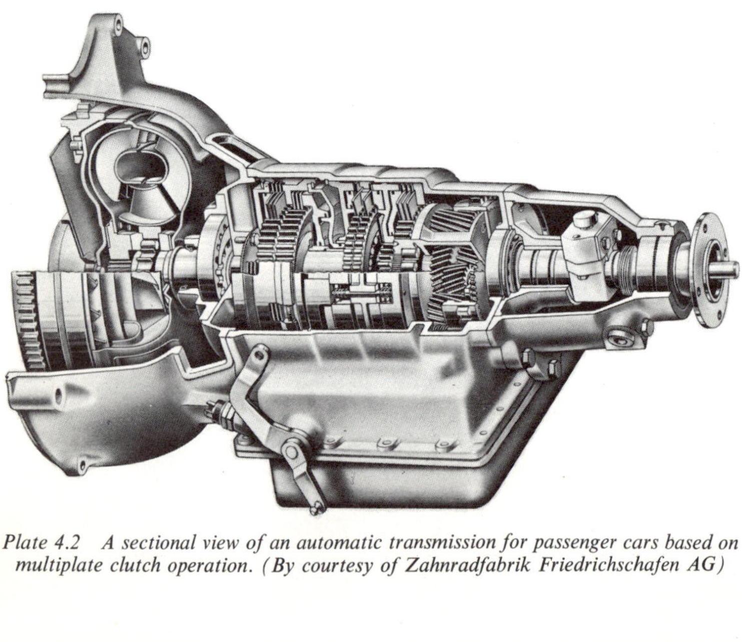

Plate 4.2 A sectional view of an automatic transmission for passenger cars based on multiplate clutch operation. (By courtesy of Zahnradfabrik Friedrichschafen AG)

surfaces are grooved to permit oil to flow through the plates, whether engaged or not, so as to dissipate the heat generated during engagement and release. In a typical design (see Fig. 4.3 and Plate 4.1) the clutch plates are assembled in a drum which locates and retains each alternate plate which is splined or has external abutments on its external diameter. The other set of plates is free of the outer drum but are retained on the clutch hub which engages with teeth or splines in the internal diameter of these plates. A piston, usually operated by hydraulic pressure, compresses the complete stack of plates against one end of the clutch drum. The mechanical load produced by the piston is transferred to each plate and each contact surface contributes some friction torque to the clutch hub and clutch drum.

In a typical assembly the clutch hub might connect or be made integral with a gear wheel, whilst the clutch drum might connect to a rotary member and would be held stationary conveniently by means of a friction band brake. Fluid under pressure must be fed, usually along the shaft, into the clutch to energise the clutch. In the well-known ZF automatic transmissions (see Plates 4.2 and 4.3) a number of such multiplate clutches are incorporated into a

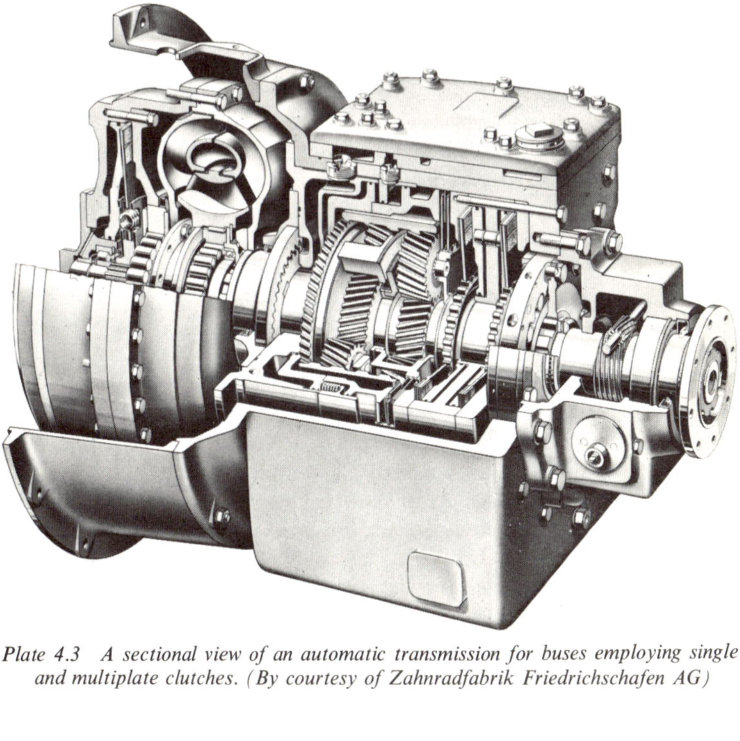

Plate 4.3 A sectional view of an automatic transmission for buses employing single and multiplate clutches. (By courtesy of Zahnradfabrik Friedrichschafen AG)

mechanical gearbox and the various gear ratios are introduced by energising the appropriate clutches attached to the required gear-wheels. This technique might be considered as a logical development of a normal gearbox resulting from the need to improve the power of the synchroniser. Having incorporated this series of clutches it becomes possible to devise a gradual release of one clutch together with a progressive application of another clutch so as to change gear without interrupting engine power. This type of automatic gearchange requires careful control to ensure that the transition from one clutch to another is effected smoothly. The presence of a torque converter between the engine and gearbox is valuable because it prevents any engine 'runaway' or smooths any gearchange 'jerk' which may result from any mistiming of the clutch transition.

4.3 EPICYCLIC AND PLANETARY GEARING

In some cases the layout of the gearbox is improved and space saved if planetary gearing is employed. In one arrangement, which is frequently included in automatic gearboxes, a multiplate clutch is combined with an epicyclic gear consisting of a sun gear, planet pinions and an annulus or ring gear. The diameter of the clutch drum is usually close to that of the annulus member and these two parts can be easily joined together or made in one piece (see Fig. 4.4). The clutch hub likewise combines conveniently with the sun gear. The multiplate clutch combines very conveniently with an epicyclic gear set of this type and the clutch may be energised to obtain a direct drive (1 :1) ratio. When released the gear set can be utilised as a normal gear ratio, and in many cases this is done by means of a friction band brake which is operated on the clutch drum or clutch hub to hold one member of the epicyclic gear stationary.

The hydraulic transmission which has been employed in Rolls-Royce and Bentley motor cars consists basically of two such combinations of gearing with a multiplate clutch and friction brake. The principles of this arrangement is shown in Plate 4.4 and illustrates how two epicyclic gear sets each combined with a friction brake and clutch are energised in different combinations to obtain four forward ratios. Table 4.1 shows the operating sequence of brakes and clutches employed in this arrangement. The front epicyclic gear set is employed to provide a reduction ratio of 2·63 : 1 or direct drive. Similarly the second epicyclic gear set provides a reduction ratio of 1·45 : 1 and a direct drive and by

Table 4.1 OPERATING SEQUENCE OF BRAKES AND CLUTCHES IN THE ROLLS-ROYCE AUTOMATIC TRANSMISSION

Gear	*Front band*	*Front clutch*	*Rear band*	*Rear clutch*	*Reverse clutch*
4th	Off	On	Off	On	Off
3rd	On	Off	Off	On	Off
2nd	Off	On	On	Off	Off
1st	On	Off	On	Off	Off
Neutral	Off	Off	Off	Off	Off
Reverse	On	Off	Off	Off	On

Table 4.2 BMC MINI TRANSMISSION GEAR TRAIN OPERATION

Selection	*Friction Bands*			*Clutches*		*Roller Clutch*
	A	D	E	F	H	B
Reverse (−2·69 : 1)	Applied	Free	Free	Applied	Free	Free
Neutral	Free	Free	Free	Free	Free	Free
First gear (2·69 : 1)	Free	Free	Free	Free	Applied	Locked
Second gear (1·845 : 1)	Free	Free	Applied	Free	Applied	Free
Third gear (1·46 : 1)	Free	Applied	Free	Free	Applied	Free
Fourth gear (1 : 1)	Free	Free	Free	Applied	Applied	Free

combining the two gear sets in series combination four forward gear ratios are obtainable:

1st gear	3·81 : 1	Reduction
2nd gear	2·63 : 1	Reduction
3rd gear	1·45 : 1	Reduction
4th gear	1 : 1	Direct

The well-known BMC Mini transmission (Fig. 4.5) manufactured by Automotive Products Ltd., represents an interesting departure from epicyclic gearing to a double bevel type twin differential gear (cf. Fig. 4.6). This is a most compact set of gearing as can be seen in the illustration and by use of three friction bands, two friction clutches and a one-way roller clutch, provides four forward gear ratios and a reverse gear to augment the torque conversion of a hydrokinetic converter-coupling.

4.4 MULTIBRAKE SYSTEMS

The friction band brake is an alternative means of energising or releasing epicyclic gear trains. The brake consists of a strip of friction material riveted or otherwise fixed to a flexible metal band

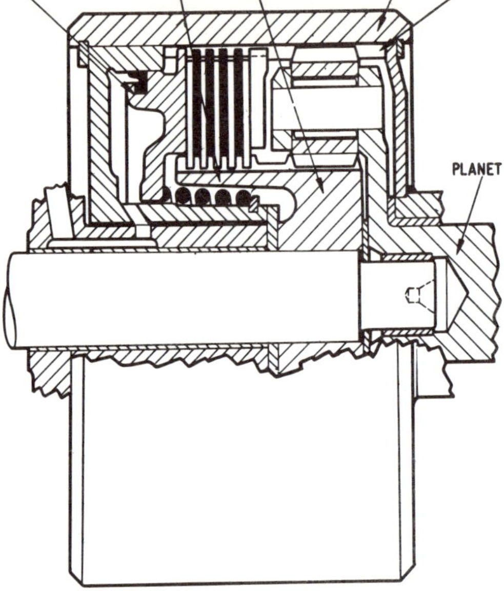

Fig. 4.4 Typical combination of a multiplate clutch with an epicyclic gear

or strap which is tightened to bring the friction material into contact with a friction drum around which the brake is mounted. The friction band has a self-wrapping tendency which permits very high braking torques to be exerted in a selected direction and friction bands are more frequently used when a large reaction torque is developed on a rotary member of an epicyclic gear train.

The friction band brake requires to be adjusted for wear if it is to retain smooth and powerful operation, and care must be taken to ensure that sufficient braking torque is available for operation under overrun conditions. If a transmission is employed for braking on steep hills and in other special conditions, a friction band brake must exert torque in the opposite direction to normal use. The self-wrapping action of the band may be lost and the brake may slip. The large forces needed to apply the band produce a resultant thrust or load on the brake drum and its supporting bearings. This load is minimised if the band can make contact with the drum for the whole circumference, and the operating linkage should preferably be balanced by a torque reacting link (see Plates 4.5 and 4.6).

The well-known Wilson gearboxes such as are employed in large

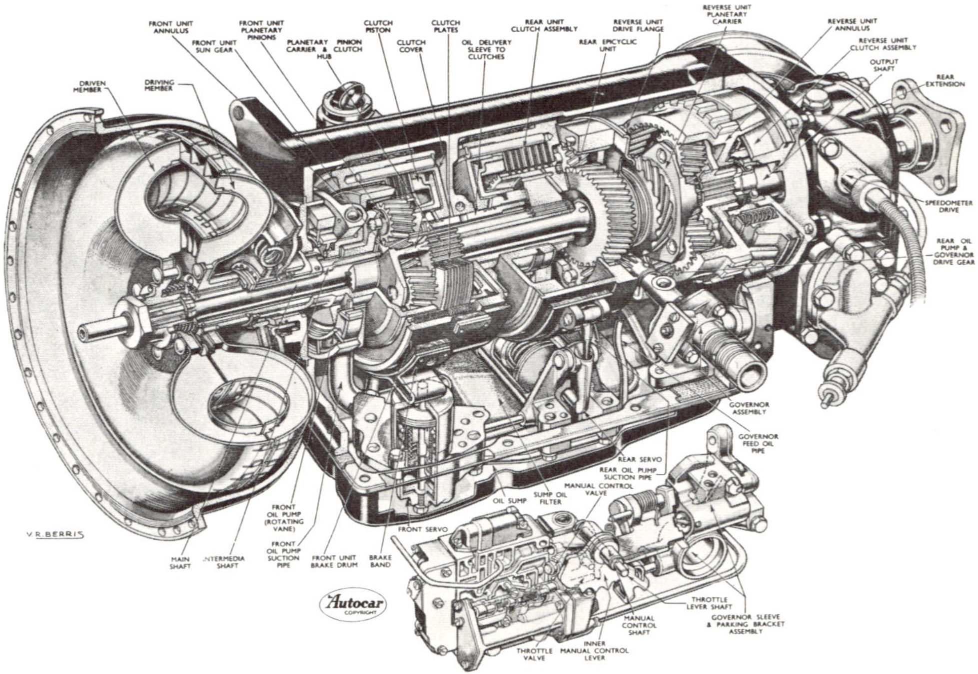

Plate 4.4 Operating sequence of friction brakes and clutches in the Rolls-Royce Automatic Transmission.
(By courtesy of 'Autocar')

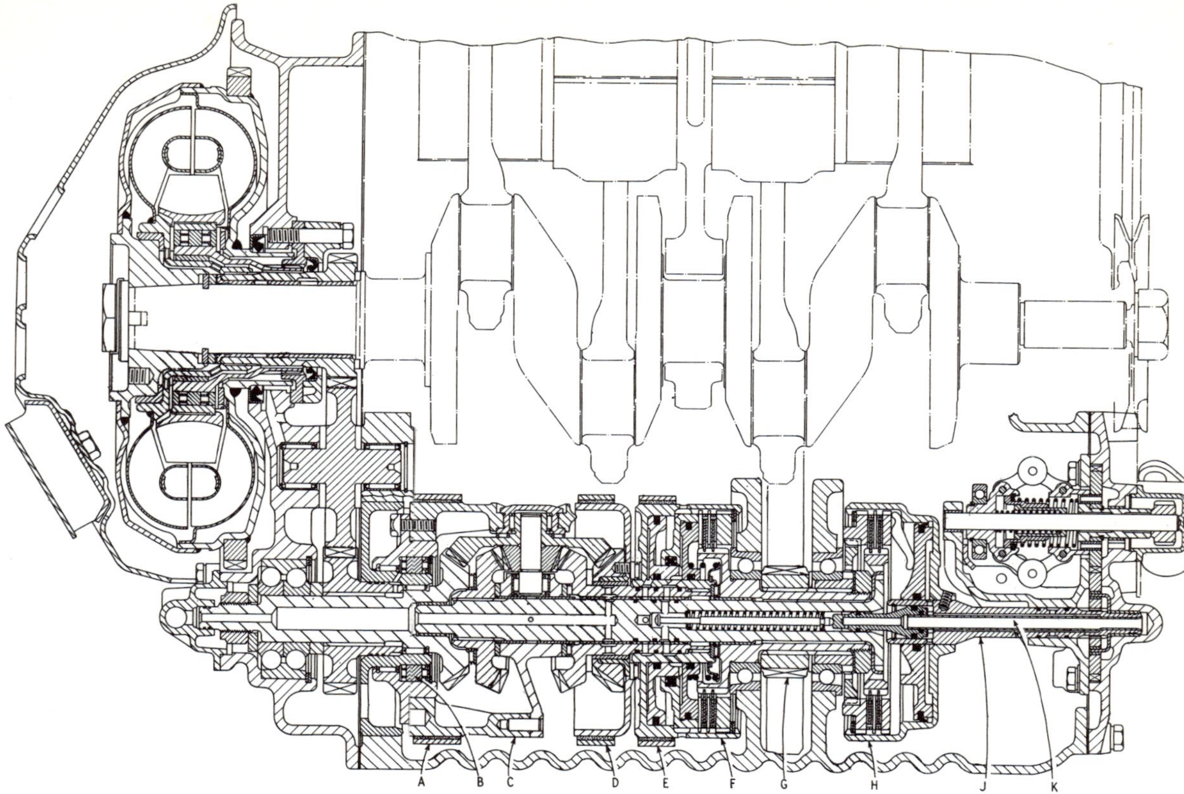

Fig. 4.5 Bevel gearing employed in the BMC–AP automatic transmission for small cars

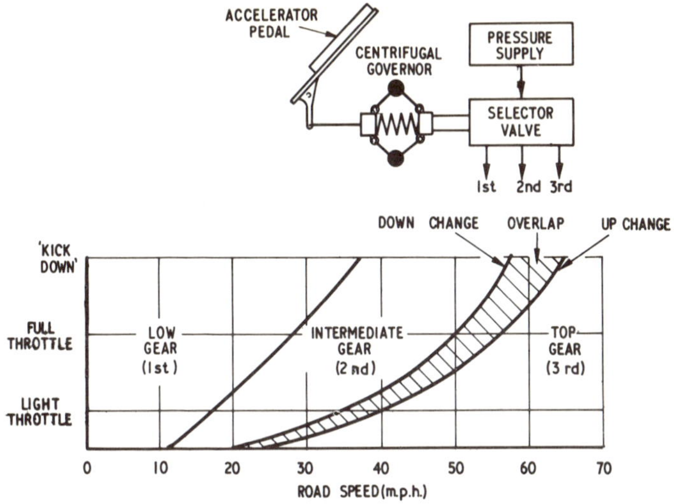

Fig. 4.6 A selector valve is moved by centrifugal governor and accelerator pedal to select gears

commercial vehicles and for higher power engines are based on the use of a highly developed friction band brake. This band is split into two strips each of which makes contact with the drum for its complete circumference. The two strips are joined together so as to make a friction band which makes two complete revolutions of the brake drum. The band tightening linkage has a high mechanical advantage so that only a light operating force is needed. In addition, an automatic adjusting screw is incorporated into each band linkage so as to take up any wear of the friction material. This type of band brake is illustrated in Plate 4.6 which shows a set of such brakes as required to operate on each of a number of epicyclic gear trains in a typical Wilson gearbox.

4.5 AUTOMATIC GEARCHANGE

The previous sections have shown various ways in which the different gear ratios are introduced in automatic gearboxes. The various clutches and brakes are energised by hydraulic, pneumatic, or electrical power supplies included in the gearbox. Whichever type of power supply is employed, however, the method of effecting a gearchange consists of a synchronised release of one clutch or brake whilst another such element is engaged. This gear change sequence is carried out whilst the engine is still

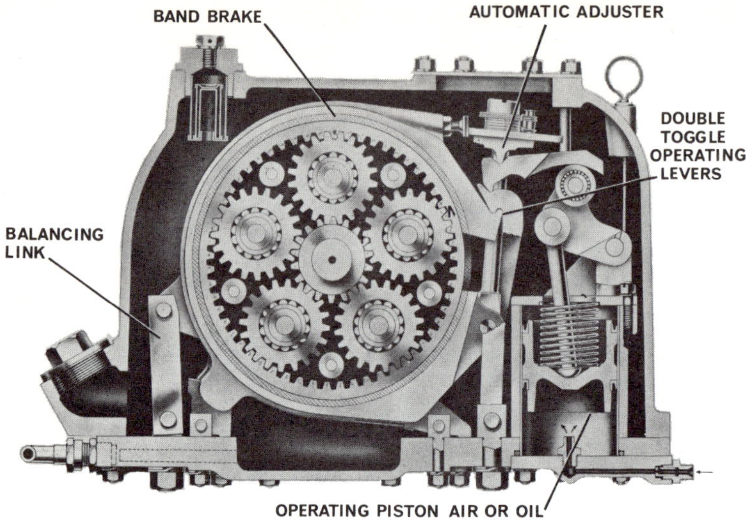

Plate 4.5 Sectional view of the Wilson friction band brake. (By courtesy of Leyland Motors Ltd.)

Plate 4.6 A view of a set of Wilson band brakes. (By courtesy of Leyland Motors Ltd.)

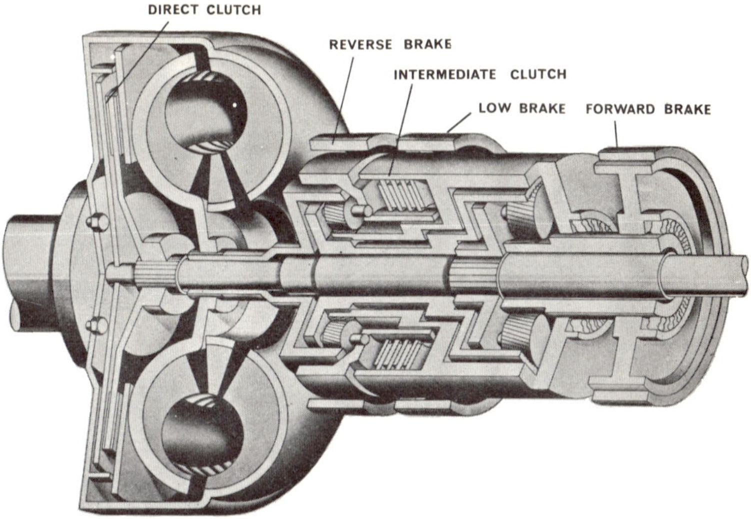

Plate 4.7 Diagrammatic view of the Borg-Warner DG Transmission, illustrating the sequence of clutch and brake operation. (By courtesy of Borg-Warner Ltd.)

delivering power and it is necessary to achieve a smooth transition from one gear ratio to another.

These powerful clutches and brakes are designed to operate at full engine torque, and if engaged suddenly would give extremely rough gearchanges, unpleasant to passengers and possibly damaging to transmission components.

It becomes necessary to instigate a short timed sequence of operation during which one friction element is gradually released whilst the next clutch or brake is applied progressively. In the case of hydraulically operated transmissions, the inclusion of small calibrated orifices or merely by correct choice of the size of hydraulic pump, it is possible to achieve a suitably timed transition from the flow of hydraulic fluid within the gearbox elements. The well-known Borg-Warner transmissions are operated by means of a hydraulic pressure obtained from pumps situated on both input and output shafts.

In the early Borg-Warner Model DG (Plate 4.7) transmission, the hydraulic circuit was comparatively simple and the gear change between top gear (1 : 1) and intermediate gear (1·45 reduction) was effected by the application or removal of hydraulic pressure to, or from, the direct drive clutch seen at the front of the transmission. The operating sequence is shown in Table 4.3.

The time taken for fluid to flow into or out of the clutch piston

Table 4.3

Selection	*Direct clutch*	*Intermediate clutch*	*Low brake*	*Forward brake*	*Reverse brake*
Reverse	Off	Off	On	Off	On
Neutral	Off	Off	Off	Off	Off
First	Off	Off	On	On	Off
Second	Off	On	Off	On	Off
Third	On	On	Off	On	Off

Table 4.4 GEARCHANGE SEQUENCE OF THE BORG-WARNER 35 TRANSMISSION

Gear selection	*Front clutch*	*Rear clutch*	*Front band*	*Rear band*	*Roller clutch* (6)
1st lock up	On	Off	Off	On	Off
1st Drive	On	Off	Off	Off	On
2nd	On	Off	On	Off	Off
3rd	On	On	Off	Off	Off
Neutral	Off	Off	Off	Off	Off
Reverse	Off	On	Off	On	Off

is just sufficient to give a satisfactory gearchange. Similarly, the gearchanges between low gear and intermediate gear are introduced by energising or release of the 'low' brake or 'intermediate' clutch. A governor valve, working from the accelerator pedal and centrifugal bobweight switches the fluid pressure for an upchange from the 'low' brake to the 'intermediate' clutch, and permits the fluid in the brake to flow through a drain orifice. On a 'downchange' this valve moves pressure to the brake and permits the fluid in the clutch to drain away.

These simple switching operations occur at different road speeds according to the acceleration pedal position and are shown diagrammatically in Fig. 4.6 which also shows an overlap of 5 m.p.h. to prevent too frequent gear changes between intermediate and top gear.

With the later Model 35 (see Fig. 4.7 and Table 4.4) it appears to be necessary to take much more care in the operation of the gearchange. It will be obvious to the reader that the 'take-up' or engagement of clutches will need to be varied according to the engine load. Thus, at full throttle, clutches and brakes need to be operated speedily and firmly, whereas at light throttle openings a slower and more gentle operation is required. Drivers of vehicles quickly learn to operate the controls on a simple manual gearbox and, probably subconsciously, to vary the clutch pedal pressure to suit the operating conditions. It is necessary to build into an

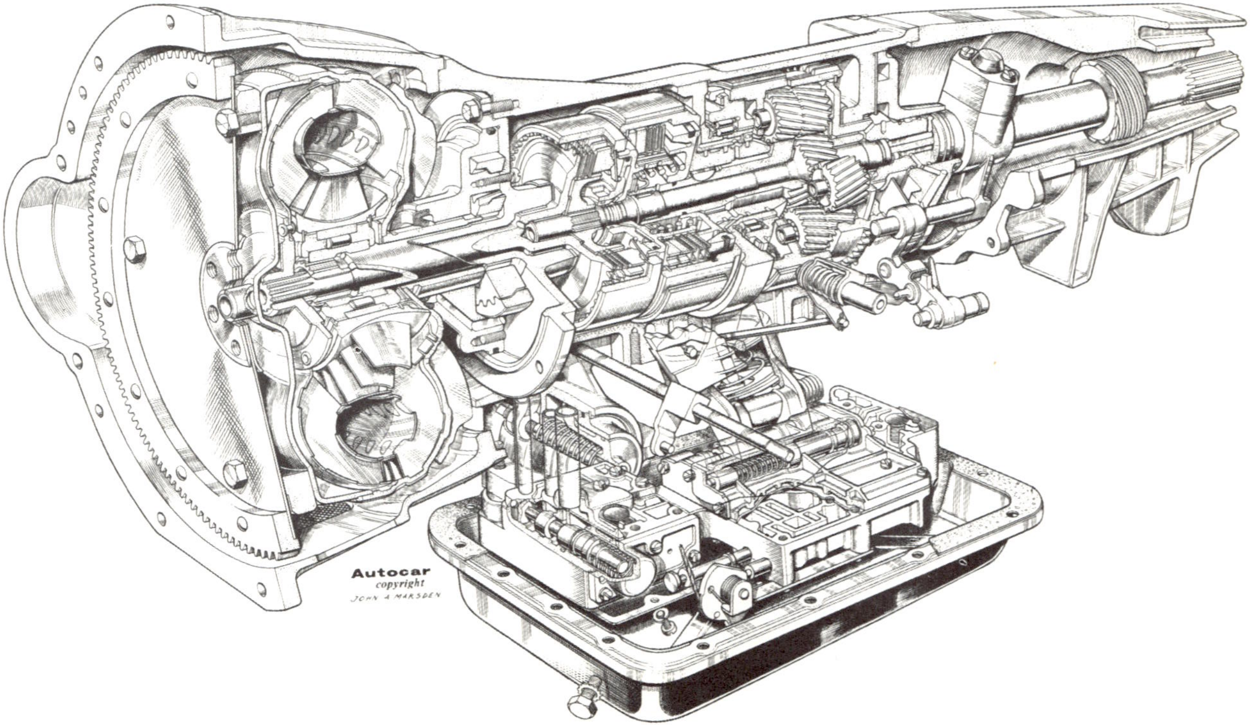

Fig. 4.7 Sectional illustration of the Borg Warner 35 transmission showing main components and gearchange sequence

automatic control system a similar intelligence to vary the speed and load of application of clutches and brakes for an automatic gearchange.

Factors which affect the load and rate of application of clutches and brakes for an automatic gearchange are the engine speed, engine torque and vehicle speed. Most automatic transmissions eventually incorporate devices which provide information or make adjustments for these conditions. It must be borne in mind, also, that hydraulic devices are sensitive to temperature and fluid viscosity changes, and that there are variations due to the wear or friction coefficient of clutches and brakes to be considered. The automatic control or brain of an automatic transmission requires a large number of valves or other components to satisfy the basic

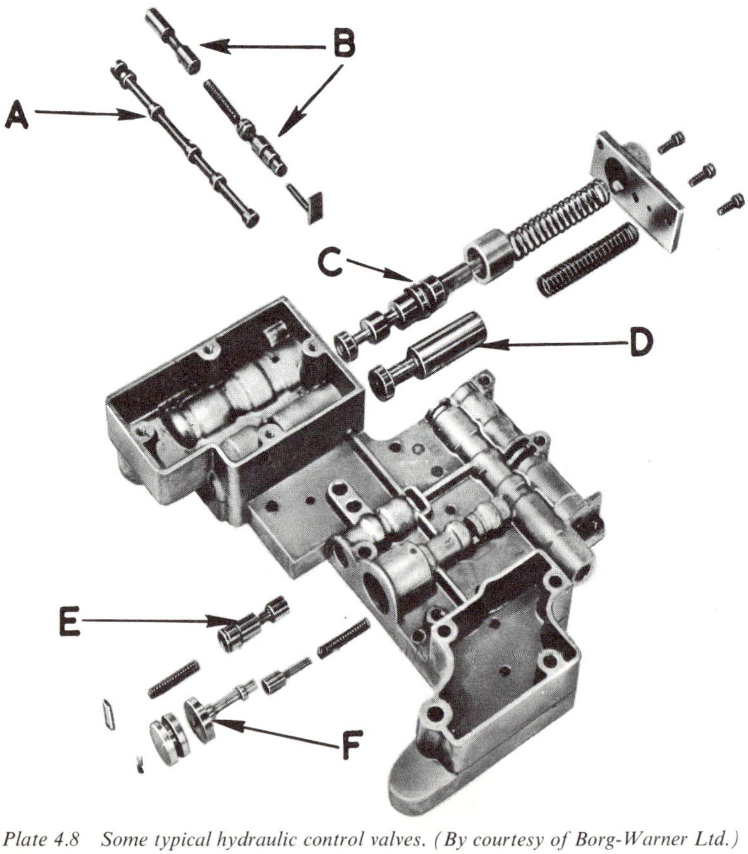

Plate 4.8 Some typical hydraulic control valves. (By courtesy of Borg-Warner Ltd.)

requirement of smooth operation and gear changes.

In the Borg-Warner Model 35 for instance, quite a number of extra valves and controls are seen to have been added to the original control circuit employed in the simpler DG hydraulic system. Some of the valves are shown in Plate 4.8.

The fluid pressure employed in the hydraulic control circuit is controlled by two regulator valves *C* and *D*. The primary regulator valve *C* is employed to supply fluid at pressures suitable for operating clutches and friction band brakes whilst the secondary regulator valve *D* supplies fluid for lubrication, to fill the torque converter etc. The actual valve is dependent on the valves *B* which vary the primary pressure according to the accelerator pedal which compresses the spring between them. These valves act as pressure relief valves and provide a 'throttle pressure' which increases as the accelerator pedal is depressed. This throttle pressure is fed back to the primary regulator valve so that the main line pressure is also increased to a higher value when the accelerator pedal is depressed. In this way the pressures employed to operate clutches and brakes are varied to suit operating conditions. At light throttle the clutches and brakes operate more gently than at higher throttle openings.

Valve *E* in this block is called the servo orifice and is employed as a restriction to delay the drop of pressure in the friction band pistons (called servos) when operating pressure is removed. This

Table 4.5

Test Point	*Name of Pressures*	*Position*	*Design Pressure*	*Notes*
1	Line	Throttle Valve	57 208 98	At idling Full Throttle Cut back
2	Throttle	Primary Regulator Valve	0 135 68	At idling Full Throttle Cut back
3	Converter	Secondary Regulator Valve	18–25	
4	Governor	Modulator Valve	0–70	According to Road speed
5	Throttle (Modulated)	Modulator Valve	0–135	Doubles throttle pressure

orifice valve permits the brakes to release slowly when a new gear is selected, so giving a smooth transfer of operation from one gear ratio to another. The modulator valve *F* is employed to vary the value of the effective pressure and again, 'throttle pressure' is reflected in the modulator valve operation. Valve *A* is a manual selector valve.

Although these valves do not represent the complete hydraulic circuit they indicate the amount of refinement needed to obtain the required degree of gear change smoothness in an automatic gearbox. In the case of the valves shown, quite accurate pressure readings need to be obtained to give the correct sequences and operating clutch loads etc. Some typical service test figures are given in Table 4.5.

4.6 TYPICAL HYDRAULIC VALVES

The valves employed in automobile control circuits are usually of the simple spool type. This is basically a cylindrical rod accurately ground so as to be a good fit in its bore. The rod is usually reduced in diameter in one or more positions so as to divide the rod into a number of lands.

In a simple manual selector valve, these lands may cover or uncover drillings into the valve bore so that different hydraulic connections are made at different positions of the valve. As seen in Fig. 4.8(a) the movement of a valve transfers a fluid supply at *A* from duct *B* to duct *C*. The fluid is retained between the two lands and flows around the valve rod to the exposed duct. By the addition of an extra land to the valve as shown in Fig. 4.8(b) it is possible to achieve a double changeover so that if a fluid supply *A* is connected to *B* then a return from *C* is connected to duct *D*. When *A* connects with *C*, then *B* is connected to *D*.

The ducts leading fluid to the valve bore are usually simple drilled holes and their position is not critical in most cases. More complex circuits can be constructed by adding more lands to the valve to perform more fluid connections.

Fig. 4.9 shows some further circuit details which can be achieved using cylindrical valves in well-fitting bores. Fig. 4.9(a) shows a relief valve using an open piston containing a spring. When the piston top is exposed to fluid pressure the spring is deflected so that some fluid is lost to a 'spill duct' until the piston load due to pressure just balances the spring load at some position between the minimum and maximum opening of the spill duct.

Fig. 4.9(b) shows how a very similar valve may be employed to obtain an approximately constant flow of fluid. A pressure supply

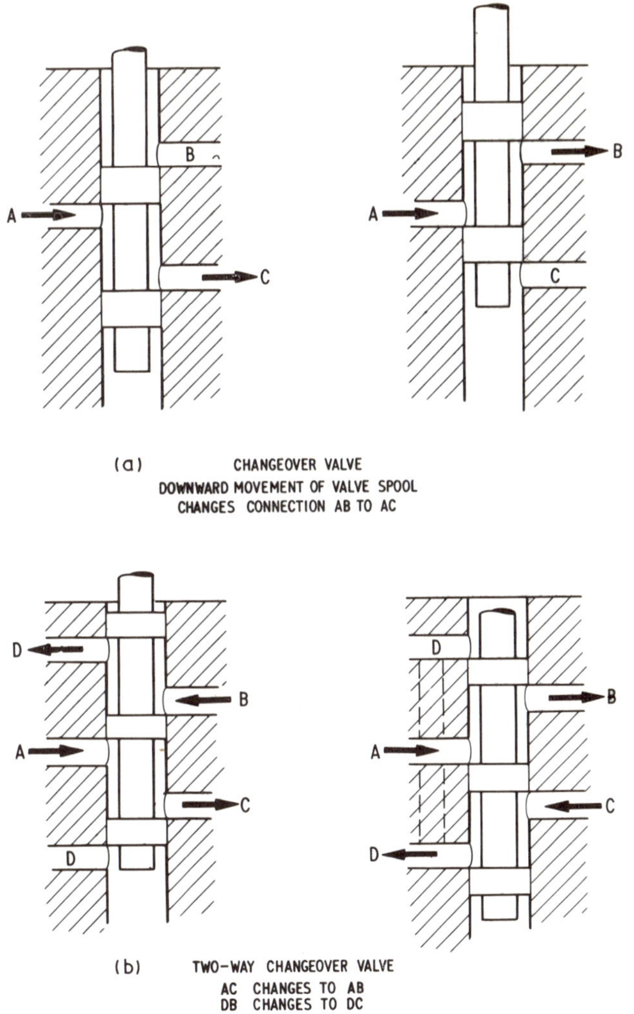

Fig. 4.8 Diagrams of typical changeover valves

flows through a calibrated orifice in the valve and out to the required supply line. The pressure developed across the orifice deflects the spring and piston downwards, partially restricting the outlet to the supply line. The valve stabilises in some position across the outlet and within a very small range of spring loads.

The pressure drop across the orifice is therefore maintained to a roughly constant value and hence the flow of fluid is maintained at a constant rate.

A piston having two different spool diameters can be employed to provide two levels of fluid pressure bearing a fixed ratio (see Fig. 4.9c). Fluid from either supply will move the valve so that a

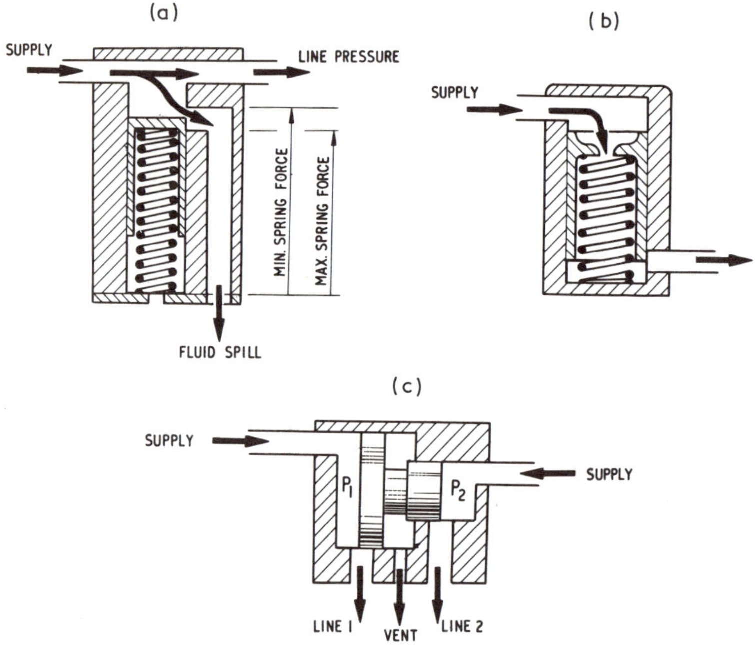

Fig. 4.9 Typical piston-type hydraulic valves: (a) Piston-type relief valve; the piston is balanced by the spring load within limits of spill duct height; (b) Constant flow valve; fluid flows through orifice in piston. Pressure on top of piston deflects spring and restricts fluid flow to a constant rate; (c) Pressure divider (modulator); piston (two diameters) moves to balance pressure loads, hence, Line 2 pressure (P_2) = 4 times Line 1 pressure (P_1)

flow to lines 1 and 2 occurs. The valve will move until the two pressure loads are equal, each piston partially restricting the outlets to lines 1 and 2 so that the pressure supply lines develop the required pressures. If the pistons are made with 2 : 1 ratio in diameter, as shown, then their respective areas will be 4 : 1 and in the balance position, when pressure loads are equal then $P_2 = 4P_1$. Each spool is acting as a relief valve and some care is needed to ensure that the position of the line 1 and line 2 ducts are accurately related to the valve dimensions.

Fig. 4.10 shows a governor valve and the method employed to obtain a pressure proportioned to output shaft speed i.e. road speed. The governor body and counter balance are fixed to the transmission output shaft. A centrifugal bob-weight is combined with a spool valve and spring mounted on the bob-weight spindle. At low shaft speeds the valve, which has two effective diameters, acts as a pressure relief valve so that line pressure in duct (2) is split beneath the valve. As the shaft speed increases, the bob-weight moves outward compressing the spring and causing the line pressure to increase. The effective diameter of the relief valve is high so that at low speeds the governor line pressures are low.

At a given shaft speed, about 500 rev/min, the spring load (increased by the centrifugal force on the bob-weight) overcomes

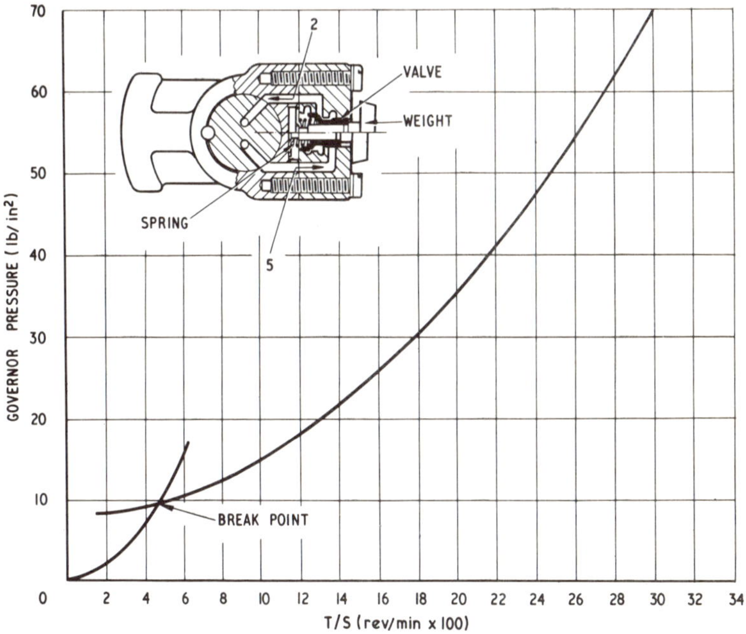

Fig. 4.10 Operation of typical governor valve. (Borg–Warner)

the pressure load and the whole assembly moves outwards so that a new fluid supply (5) and a new smaller diameter of the valve becomes effective. The centrifugal force of the bob-weight opposes the governor pressure giving rise to a governor line pressure as shown on the graph. The combined characteristic gives better control at lower shaft speeds when the low gears are engaged. Also, below the 'break point' clutch pressures can be increased.

Fig. 4.11 shows how this governor pressure could be applied against throttle pressure to operate a 'shift' valve. The ends of this valve are exposed to governor pressure and throttle pressure respectively, and the valve will move from the higher pressure to the side at lower pressure. Movement of the shift valve from one side to another changes the fluid pressure from one circuit to another. In this example high governor pressure (high road speeds) moves the valve to select top gear, whereas high throttle pressure tends to move the valve to obtain a lower gear.

Comparison of the two graphs shows that at a low throttle opening governor pressure exceeds throttle pressure at about 32 m.p.h., but if the accelerator pedal is depressed to its full throttle position the road speed must increase to approx. 60 m.p.h. before the governor pressure exceeds throttle pressure and selects top gear. By use of extra pedal movement it is quite common to have a 'kickdown' enabling a downchange to a lower gear to be obtained over a wide range of road speeds.

In this example the shift valve is seen to direct a supply of pressure fluid to energise either a friction clutch or a band brake to obtain the desired gear ratio. In modern transmissions, the 'shift valve' provides a simple 'switching operation' starting a timed sequence to release the band and energise the clutch (or vice versa).

Thus there would be some restrictor in the supply line to each clutch or band so that the new member was engaged rapidly but the old member was released more slowly. In some cases a 'feed-back line' might delay the release of the old member until the new member had started to engage.

These refinements all require the addition of further pistons, valves etc. so that these hydraulic control systems eventually become more complex for design and manufacture than the gear train and actual power transmitting components. Some idea of the number of valves and hydraulic controls in an automatic transmission can be seen in Plate 4.9 which shows the valve block items for the Vauxhall Hydramatic transmission which became available in 1960

4.7 DRIVER'S OVER-RIDE AND SPECIAL OPERATING CONDITIONS

So far consideration has merely been given to automatic selection of gears to suit typical road operation. There are many occasions

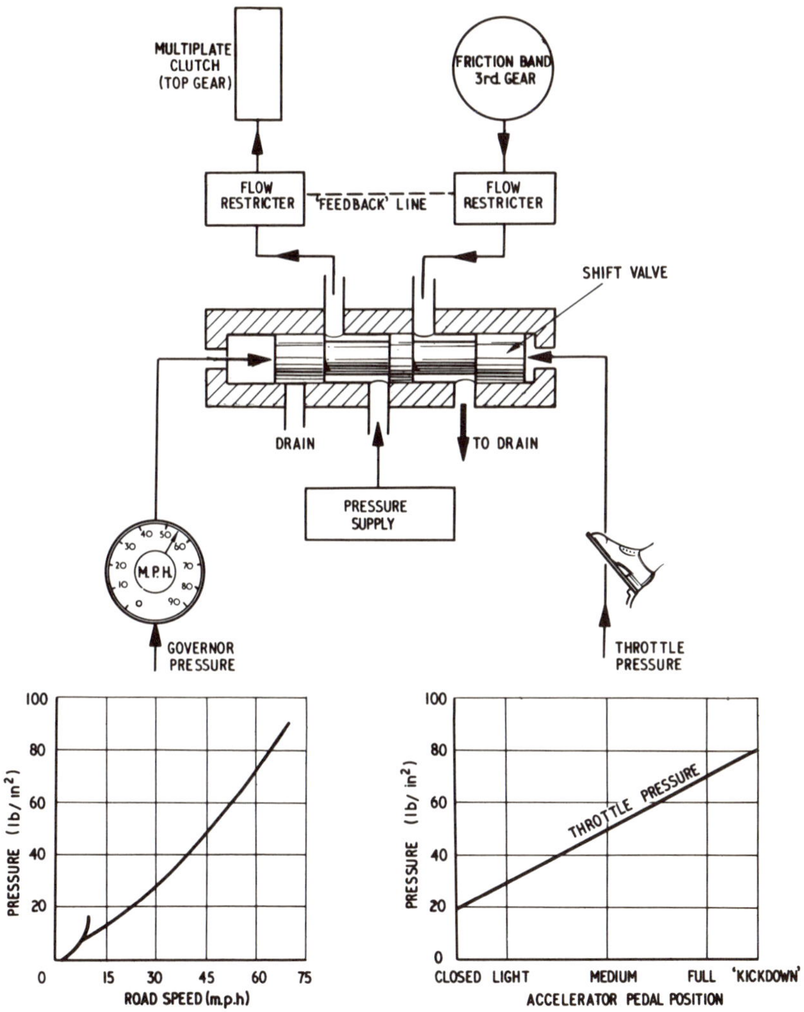

Fig.4.11 Diagrammatic operation of a 'shift' valve to change from a brake operation to a clutch engagement

when a driver, due to traffic or special road surface conditions, ice, mud etc., will prefer to stay in one particular gear ratio. It will be necessary to provide a means for the driver to over-ride the automatic sequence and substitute his own choice. This in itself brings a further train of requirements to ensure that the engine and vehicle are protected against an accidental operation of its own

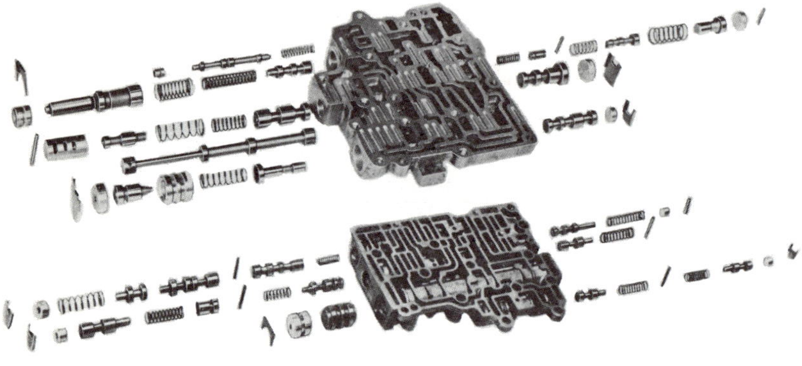

Plate 4.9 Exploded view of a hydraulic control block assembly. (By courtesy of Vauxhall Motors Ltd.)

controls. The accidental selection of reverse whilst travelling at speed in the forward direction is a typical example.

Thus, in addition to carrying out the design functions, an automatic transmission must include some further duties which result from its own particular properties. Some of these cxtra duties and components are briefly considered below.

4.7.1 MANUAL SELECTOR

There are certain road conditions and special circumstances, such as the use of an engine to assist braking on steep descents, when the automatic gear-change sequence is not suitable. It is necessary to provide a manual gear lever or selector so that, when required, a driver may make his own choice of gear ratio to be employed at any particular time. It is also necessary to have this manual lever to select forward, neutral and reverse which are decisions that can only be made by a driver. Some selectors also include a 'park' selection which involves the application of a parking brake. (See Plate 4.10).

In commercial vehicles a simple manual lever 'inches' the

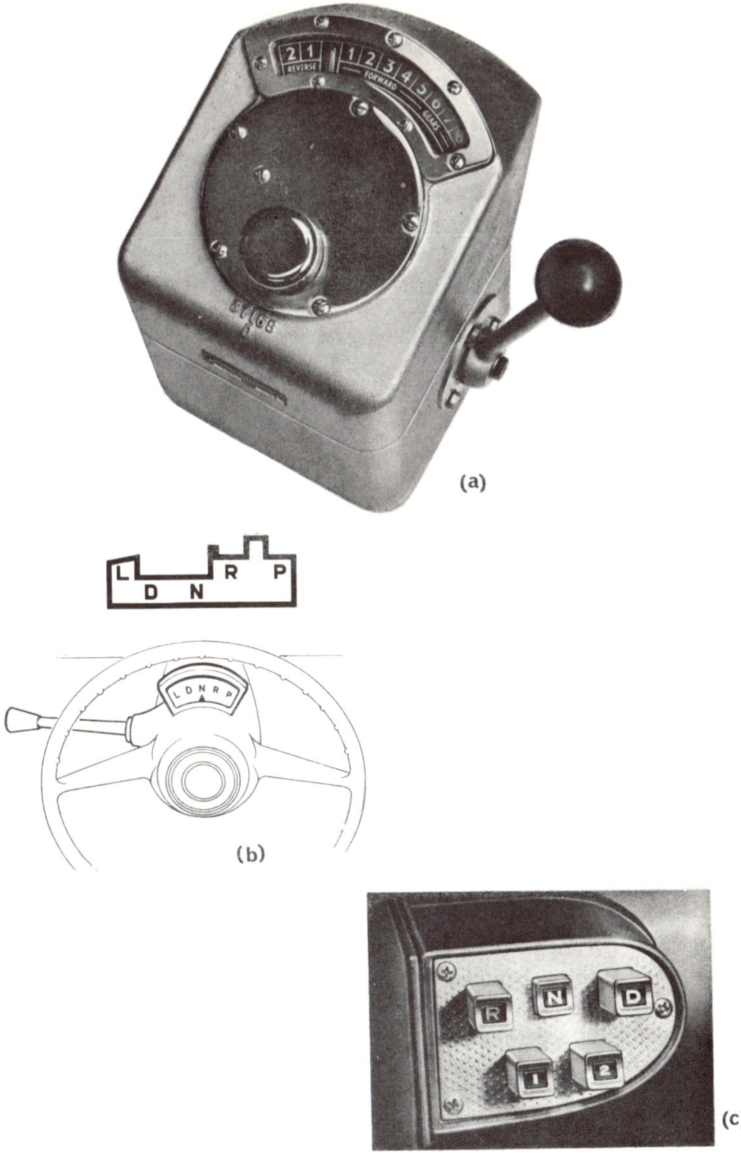

Plate 4.10 Typical manual selectors for automatic transmissions.
(a) An 'inching' lever for commercial vehicles. (By courtesy of Leyland Motors Ltd.)
(b) Steering column selector for passenger cars. (By courtesy of Borg-Warner Ltd.)
(c) Push-button selector. (By courtesy of Chrysler Motors Ltd.)

controller up and down the available range one step at a time, but in passenger cars the manual selectors are less frequently used.

4.7.2 TOW-STARTING

Most transmissions incorporate some means of starting a vehicle's engine by towing or running down a gradient. This is the main reason for the inclusion of a rear pump driven via the axle and transmission output shaft. Automatic transmissions will develop pressure from the rear pump as soon as the vehicle moves and this pressure can be employed to energise the various brakes, clutches or other controls in the transmission. Although the transmission's front pump is normally employed as the main hydraulic power supply, this pump will not rotate until the engine fires. Tow starting can be achieved by using the rear pump to supply the necessary fluid and pressure. A simple changeover valve is sufficient in some transmissions, and pressure from the front pump will then return the tow start valve to its normal position.

4.7.3 HUNTING

One common problem with automatic transmissions incorporating gearchanges is called hunting. Having selected an automatic control which introduces a gearchange at say 30 m.p.h., when the vehicle is cruising at that speed the automatic control or governor may proceed to change up and down continuously. It is necessary to introduce some 'hysteresis' or a lag in the control so that if the 'upchange' takes place at 30 m.p.h., then a downchange will not occur until there has been a significant reduction in road speed to 28 m.p.h. or perhaps 25 m.p.h., at the same throttle setting. Most governors or controllers operate against a mechanical spring, so that it is common practise for the spring load to be adjusted by a hydraulic valve, so that this load is reduced after an upchange and increased after a downchange. This effect was shown in Fig. 4.6.

The governor valve shown in Fig. 4.12 achieves this hysteresis by a small spring loaded plunger mounted at right angles to the governor valve and acting on a series of ramp profiles. In this way a slight extra mechanical load is required to move the governor valve to each new position. As a result, a similar effect is achieved, a reduction in spring force for downchanges and an increase for upchanges. This small plunger not only separates the upchange and downchange speeds but helps to give a positive movement of the governor valve from one position to another. This particular gearchange system was employed in the Hobbs 'Mechamatic' transmission, and road speed was measured by the pressure from a gear pump which moves the valve downwards against springs. Various bleed valves in the 'kickdown' valve were used to adjust this pressure depending on the accelerator pedal position.

4.7.4 THROTTLE SENSITIVITY

In previous paragraphs, it has been mentioned that the road speed at which a particular gearchange occurs must vary according to

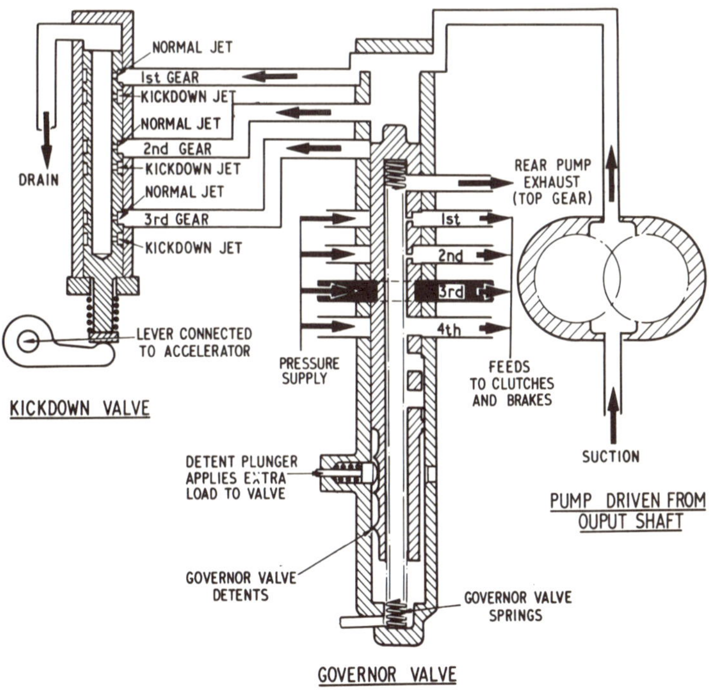

Fig. 4.12 A detent plunger applies additional load to a governor valve to prevent rapid fluctuation and hence, repeated gearchanges

gradient, traffic conditions and the driver's wishes etc. All these factors can be taken into account by noting the position of the accelerator pedal position, or alternatively, the throttle butterfly valve position.

Whereas a driver may wish to be in top gear by 20 m.p.h., in certain conditions, when his journey is more urgent, he may wish to delay his gearchange into a top gear to a speed in excess of 60 m.p.h. Consequently, most governors or controllers must be capable of adjusting the gearchange speed over such limits. In the accompanying graphs it will be noted however that, depending on

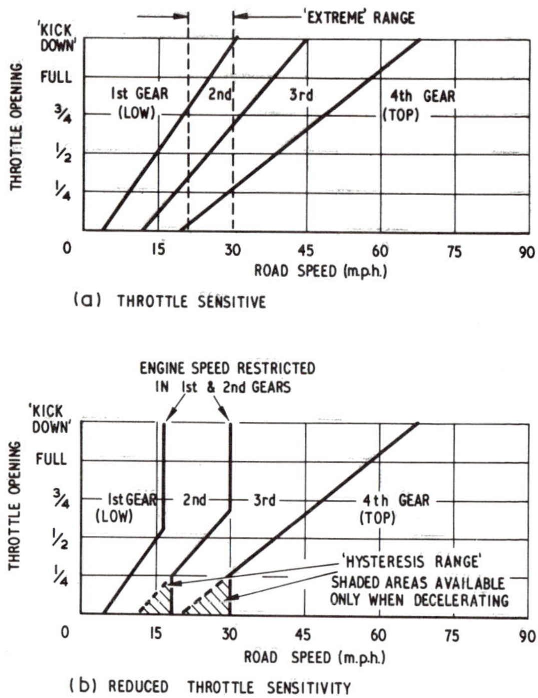

Fig. 4.13 Throttle sensitivity may be reduced by restricting engine speed in low gears and by widening the hysteresis in range

the ratio step between each gear, there can be considerable overlapping of possible gear ratios at certain road speeds.

In Fig. 4.13a it will be noted that a vehicle's transmission can be governed so that top gear is available from 20 m.p.h. to its maximum speed of 90 m.p.h. If however the same technique is applicable to all other available gear ratios, it will be noted that between 20 and 30 m.p.h., any of the four gear ratios is obtainable depending on the position of the accelerator pedal. In this example the vehicle/transmission combination is extremely throttle sensitive.

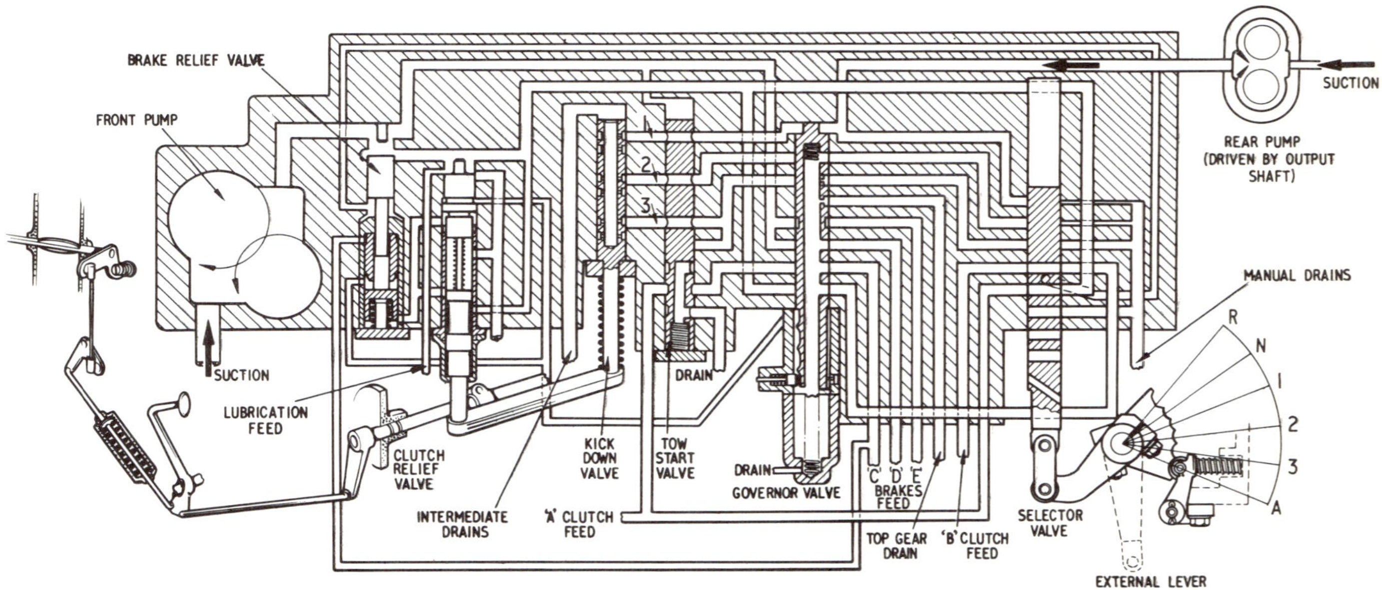

Fig. 4.14 A complete hydraulic circuit diagram. (Hobbs)

Most vehicle operation would be satisfactory if only one gear-change could be obtained at any road speed, and as shown in Fig. 4.13b a more satisfactory arrangement is obtained by making the lower gears less dependent on the throttle position and by cutting off the use of extremely high rev/min in lower gears. Throttle sensitivity is improved if a wider range of hysteresis is employed to increase speed difference between 'upchange' and 'downchange' speeds.

4.7.5 SAFETY DEVICES

The provision of automatic gearchanges, together with the use of a manual selector makes it necessary to prevent various types of mishandling which can be easily made by careless operation of the selector lever. Although similar mishandling can be made with an ordinary synchromesh gearbox, the driver is usually warned by the crashing of gears that some error has been made. Improper selection of Low, Reverse, etc., is easily made with an automatic transmission which contains sufficient power to execute the selection, but the effect on the engine, vehicle or passengers may be disastrous if the selection is carried out at high road speed. Most transmissions contain certain safety devices—usually working from the rear pump—to ensure that dangerous selections or combinations are prevented (inhibited). The safety of engine or vehicle due to component failure must also be considered.

This chapter has shown that the automatic operation of gear-boxes or transmissions has not only introduced new types of components for power transmission but has required the provision of a 'brain' or automatic control to replace the driver's decision-taking and manual operations. Much of the cost and complication of the new transmissions appears in the components and techniques needed to reproduce the driver's skill and judgement. Summarising the duties or functions of the automatic control system, it will be recognised that the following functions at least must be provided by the controller for a stepped ratio transmission.

(1) Selection of gear ratio—dependent on engine load and road speed.

(2) Operation of clutches/brakes etc.:
  - (a) maximum operating load dependent on engine load,
  - (b) rate of application dependent on engine speed and load,
  - (c) timed sequence for transition from clutch to another manual selector and special selections.

(3) Provision for tow-starting.

Plate 4.11 Sectional view of the Hobbs Mechamatic transmission shows relative size and space occupied by the hydraulic components as compared to gearing. (By courtesy of Hobbs Bros. Ltd.)

(4) Prevention of hunting, throttle sensitivity.
(5) Safety precautions.

From the above list it is easy to see how the control circuit grows in components and complexity. There is clearly something to be gained in efforts to simplify these circuits, but the Hobbs transmission which had a comparatively simple hydraulic circuit (see Fig. 4.14) is complex enough for most engineers. Other engineers might argue that in order to ensure smooth operation of a gear-change it is necessary to consider each gear ratio separately and hence the large collection of shift valves, throttle valves and other components in the more refined circuits. A sectional view of this transmission is shown in Plate 4.11 and illustrates what a relatively small proportion of the transmission is given to the gear train and emphasises that the major difficulty with a mechanical gearbox is still in making the transition from one gear ratio to another. Although this problem may be solved in the future, it seems possible that stepped gearboxes will always present these difficulties and that some form of 'stepless' gearbox or drive will be more satisfactory. The current automatic gearboxes appear to have been overwhelmed by the hydraulic components.

BIBLIOGRAPHY

1. YOUNGS, P. R., 'The Transition from Torque Converter to Fluid Coupling', *Trans. Soc. Automob. Engrs.*, **6**, No. 1, 134
2. EKSERGIAN, R., 'Torque Converter and Fluid Coupling', *J. Franklin Inst.*, **235**, No. 5, 441
3. ATKINS, A. J., 'An Automatic Transmission for Small Cars', *Proc. Inst. Mech. Engrs. (London) Automob. Div.*, **181**, Pt. 2a, No. 6, 173 (April 1967)
4. ANON., 'Borg-Warner Model 35—Automatic Transmission', *Autocar*, 24–28 (30 Mar., 1967)
5. ANSDALE, R. F., 'Transmissions—Band Operation 3-speed Automatic Transmission', *Automob. Engr.*, **57**, No. 6, 214/5 (May, 1967)
6. DÖRPMUND, H. and SOMMER, O., 'The Fully Automatic Transmission of the VW 1600', *A.T.Z.*, **69**, No. 9, 285–8 (Sept., 1967)
7. ANON., 'Hobbs Mecha-matic Transmission', *Automob. Engr.*, **49**, No. 3, 83 (Mar., 1959)
8. ANON., 'Voith Automatic Transmission for Cars', *Automob. Engr.*, **55**, No. 11, 435–6 (Oct., 1965)
9. MCNEIL, I., *Hydraulic Operation and Control of Machines*. Thames & Hudson (1955)
10. ANON., 'Automatic Transmission Fluids', *Automob. Engr.*, **57**, No. 2, 53–4 (Feb., 1967)
11. LUDWIG, L. P., 'Engagement Characteristics of Wet-Type Clutches', *S.A.E. Trans.*, **67**, 391 (1959)

5

overdrives and economy

5.1 INTRODUCTION

The continued popularity and success of various overdrives available on a wide range of vehicles is an interesting and perhaps surprising feature of vehicle transmission equipment. An overdrive consists of a small auxiliary gearbox, usually mounted behind a vehicle's main multi-speed gearbox, arranged to provide a speed-increasing gear ratio which can be used in conjunction with the higher gears in the main gearbox (see Plate 5.1). As its name implies, the overdrive enables the transmission shaft to be over-driven at a higher speed than the engine speed.

The reason for the adoption of this extra gearbox is to provide a special ratio for cruising when the engine can provide the necessary power at a lower output speed and at a reduced fuel consumption. Also, if the main gearbox ratios are widely spaced, the overdrive can supplement it by offering a choice of extra gear ratios for higher road speeds.

The surprising aspect of these overdrives has been the fact that vehicle manufacturers have not increased the number of ratios in the gearbox to offer the same facility, together with the fact that comparatively large numbers of overdrives are purchased by owners at a relatively high extra cost. There are of course exceptions to the above, and certain manufacturers have offered gearboxes which include additional ratios or axle combinations to provide 'overdrive' operation. In general, however, drivers appear to prefer to flick a simple lever or switch to bring the overdrive ratio in and out of operation rather than to carry out the normal gear-change operation.

Thus, although overdrives are frequently offered in passenger cars as an alternative to an automatic transmission, part of the reason for their popularity can be traced to the same desire for simpler and easier gearbox operation that has produced automatic transmissions. In the case of an overdrive, the additional extra cost

Plate 5.1 A typical arrangement of an overdrive unit added to a passenger car gearbox. (By courtesy of 'Automobile Engineer')

will be wholly or, at least partly, absorbed in savings in fuel achieved by its use. Such saving depends on proportion of time for which the overdrive is used and the type of driving, but about 10–15% reduction in fuel costs seems possible.

On the other hand, the use of automatic transmissions tends to show a slight overall increase in fuel costs, but several independent tests have shown that certain automatic transmissions bring a similar 10–15% reduction of fuel costs for city use and in heavy traffic conditions. It would appear that some form of automatic transmission which includes an overdrive, or overdrive ratios, will provide the final solution and will obtain the required ease and simplicity of operation without the disadvantage of increased fuel costs.

The Roverdrive transmission which was employed for a period in about 1956 on Rover cars was an attempt to combine an automatic transmission with an overdrive gear. This result was achieved by fitting a conventional overdrive unit at the end of a fluid converter and auxiliary gearbox—making it a rather bulky and complicated unit. The incentive remains, however, to produce an automatic transmission with overdrive ratios but a simple and cheap mechanism is needed and some of the devices shown in Chapter 6 may eventually prove satisfactory on this score.

There can be no doubt however that an overdrive gearbox offers advantages in operating economy. This economy is most easily shown in passenger cars where, in order to maintain an adequate reserve of power for hill climbing or acceleration, the normal 'top gear' is somewhat lower than might otherwise be chosen.

5.2 POWER REQUIREMENT

As shown in Fig. 5.1 the power available from an engine in a normal top gear is usually arranged to be considerably higher at each road speed than that required to overcome the various air and rolling resistances, as shown by the shaded area (level ground). This power curve meets the resistance curve at approx. 90 m.p.h. and slightly below the maximum point of the power curve. This

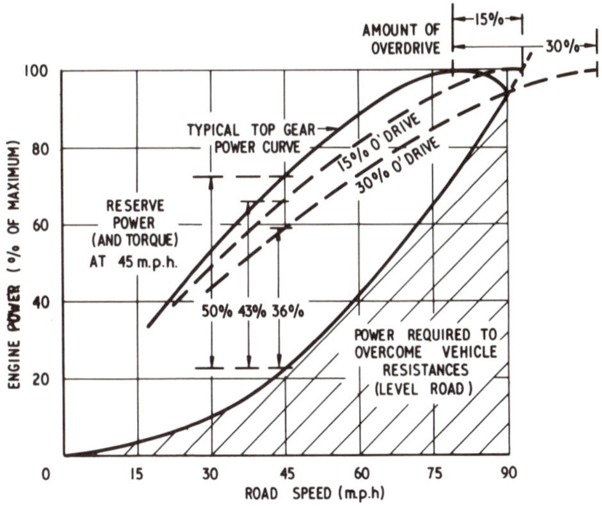

Fig. 5.1 Comparison of power available with that required to drive the vehicle at various road speeds using normal 'top' gear and overdrive ratios

typical choice of 'gear' (i.e. wheel size and axle reduction combined) allows an adequate reserve for acceleration and hill climbing and for possible variations or deteriorations of engine power. As a result, a few m.p.h. in maximum speed is lost, because the peak power is equivalent to the total losses at approx. 92 m.p.h.

By increasing the top gear by about 15%, the power curve could be made to intersect the loss curve at the point of maximum power so as to achieve the maximum level road speed of the vehicle. There would be a reduction of the reserve power available at all road speeds. By increasing the overdrive ratio to 30% above top gear it is seen that the point of intersection of power available with the power requirement curve is brought back to 90 m.p.h. and a further reduction in reserves is made.

The reason for suggesting such overdrive ratios is in the reduction of engine speed which they achieve at all road speeds. Reduced engine speeds means less noise inside the vehicle, less wear, and more important, reduced fuel consumption. The penalty of these higher gears is seen by the reduction in torque or power reserve which is small at half maximum speed (45 m.p.h.) but is more pronounced at higher road speeds. Increased passenger load, slight gradient, or high winds can quickly absorb this reduced reserve and a driver would need to change back into top gear for any of these reasons. The disadvantage of frequent gear changing is minimised by automatic operation using a light switch or lever. This type of gearchange is often termed 'hot-shift' or semi-automatic because there is no interruption of engine power, but requires the driver to initiate the change.

5.3 FUEL ECONOMY

The important fuel saving properties of an overdrive arise from the fuel consumption characteristics of internal combustion engines which usually have reduced economy at low throttle or low torque settings. Overdrive gears reduce engine speed at each road speed as compared to a typical top gear so that the engine operates at a slower speed and higher torque when the overdrive is engaged. Engine operation at low speed and high torque is always more economical than when the same output power is obtained at higher speed and lower torque settings, because higher engine speeds bring higher friction and other engine losses.

The principle of overdrive operation can be demonstrated by a study of Fig. 5.2 which shows a typical $1\frac{1}{2}$ litre engine power curve on which has been superimposed its fuel consumption curves (dotted) indicating the specific fuel consumption in lb/bhp h. The curves show how the efficiency of the engine varies according to the output power and engine speed. The maximum efficiency of the engine is obtained at about 25–30 hp and between speeds

2,000–2,750 rev/min. At any other conditions the efficiency falls and higher specific fuel consumptions are shown.

Scales at the top of the graph relate engine speed to road speed in a typical top gear and in an overdrive gear. It will be noted that a maximum speed of 90 m.p.h. is equivalent to 4,500 rev/min in top

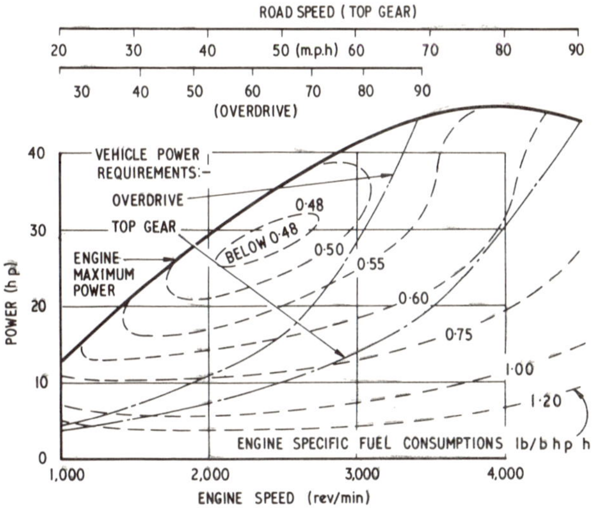

Fig. 5.2 Top gear and overdrive gear power requirement curves compared against engine specific fuel consumptions

gear but to only 3,400 rev/min in an overdrive gear (30%). The power requirement curve of the vehicle has been plotted in relation to both these gear ratios and it is possible to compare the specific fuel consumption with either gear in operation at any road speed.

Comparing for instance the maximum speed of 90 m.p.h. representing about 44 bhp at the road wheels, then in top gear the specific fuel consumption is (by interpolation) approx. 0·65 lb/bhp h whereas in overdrive the figure is 0·53 lb/bhp h. Thus in one hour at 90 m.p.h. the vehicle would consume 28·6 lb (3·8 gal) of fuel in top gear but in overdrive would consume only 23·3 lb (3·1 gal) i.e. a saving of 0·7 gal in 3·8 gal (18%).

Similarly by comparing the power requirements over a range of speeds, the saving obtained by the overdrive gear can be predicted. The curves shown are typical of an economical $1\frac{1}{2}$ litre engine but most engines will show similar distributions and gains with an overdrive.

The estimated savings in fuel as obtained from Fig. 5.2 are shown in Table 5.1.

Table 5.1 CONSTANT SPEED FUEL CONSUMPTION

	Top Gear					*Overdrive*				
Road Speed (m.p.h.)	*Power requirement* (bhp)	*Specific fuel consumption*	*Fuel consumption* (lb/h)	(gal/h)	m.p.g.	*Specific fuel consumption*	*Fuel consumption* (lb/h)	(gal/h)	m.p.g.	*Saving* %
30	5	1·00	5·00	0·668	44·6	1·00	5·0	0·668	44·6	0
40	7·5	0·90	6·75	0·900	44·4	0·87	6·5	0·87	46·0	4
50	10	0·80	8·0	1·07	46·7	0·76	7·6	1·02	49·0	5
60	14	0·72	10·1	1·35	44·3	0·63	8·8	1·18	51·0	15
70	20	0·65	13·0	1·74	40·3	0·56	11·2	1·50	46·9	16
80	30·5	0·61	18·6	2·50	32·0	0·51	15·6	2·09	38·2	19
90	44	0·65	28·6	3·83	23·0	0·53	23·3	3·10	29·0	22

Table 5.2 FULL THROTTLE FUEL CONSUMPTION

	Top Gear					*Overdrive*					
Road Speed (m.p.h.)	*Maximum Available Power* (bhp)	*Specific Fuel Consumption* (lb/bhp h)	*Fuel consumption* (lb/h)	(gal/h)	m.p.g.	*Maximum Available Power* (bhp)	*Specific Fuel Consumption* (lb/bhp h)	*Fuel consumption* (lb/h)	(gal/h)	m.p.g.	*Saving* %
30	21·5	0·52	11·2	1·50	20	15·5	0·57	8·9	1·15	26·1	26
40	29	0·49	14·2	1·90	21·1	22·4	0·52	11·6	1·55	25·8	22
50	35·5	0·49	17·4	2·32	21·5	27·5	0·49	13·5	1·80	27·6	29
60	41·0	0·51	21·0	2·80	21·5	33·0	0·49	16·2	2·15	28·0	30
70	45·0	0·53	23·8	3·19	22·0	38·0	0·50	19·0	2·54	27·5	25
80	46	0·56	25·8	3·45	23·1	41·7	0·51	21·3	2·84	28·2	22
90	44	0·65	28·6	3·83	23·5	44·0	0·53	23·3	3·11	28·9	22

Vehicle fuel consumptions at a steady constant speed are always more favourable than the overall fuel consumption of a vehicle which includes low gear use, hill-climbing and acceleration, etc. In the case of a top gear where a steady speed is obtained at 50% of engine torque then the fuel consumption would be doubled if the engine is working at full torque at that road speed. An overdrive gear also shows a gain for full throttle operation and a comparison of fuel consumptions in top gear and in overdrive gear at various road speeds given in Table 5.2.

It will be noted that there is an improvement of over 20% over this speed range by use of the overdrive gear. These calculations make no allowance for any gearing losses when the overdrive is introduced so that some reduction in practice must be expected on these figures. However it is clear that a useful reduction in fuel consumption may be expected. The overall saving is some mean value between the full throttle and steady speed gains.

This gain in economy obtained by overdrive gears is an exploitation of the engine's operating efficiency. A higher 'step-up' ratio would bring a further gain but with a corresponding increase in gearchanging, and it is not usual to exceed an overdrive of 30% in passenger cars.

Diesel engines employed in commercial vehicles are more efficient than carburetted engines but show a similar effect when overdrive ratios are employed. It is not usual to have a separate gearbox for the overdrive performance probably because the improvement obtained is not so great and a 30% step is rather large by commercial vehicle standards. The road resistances of commercial vehicles are much more dependent on the vehicle's load and, as mentioned in the previous chapter, it is necessary to have a large number of more closely spaced gear ratios to ensure economical operation under all conditions.

In some cases it is possible to introduce a fairly large step-up ratio by means of a two speed axle so that the unladen vehicle can make a return journey at higher speed and more economically. Fig. 5.3 shows similar constant fuel consumption loops when plotted on the torque curve of a large 6-cylinder diesel engine. This, again, is a very efficient engine and smaller or higher speed engines will be expected to show higher specific fuel consumptions.

A typical choice of a 'top gear' would leave a reserve of about half engine torque when the laden vehicle is travelling on a level road. In an unladen condition the vehicle resistances are likely to be less than half the laden values so that in the unladen condition the engine will be working at very low load and reduced efficiency. A 30% overdrive ratio obtained by an axle ratio or other means is seen on the graph to lift both torque requirement curves into reduced fuel consumption areas, but the amount of gain is much more noticeable for the unladen condition. At 40 m.p.h. in the unladen condition, the necessary power is obtained at a specific fuel consumption of 0·56 lb/bhp h in top gear but at only 0·43

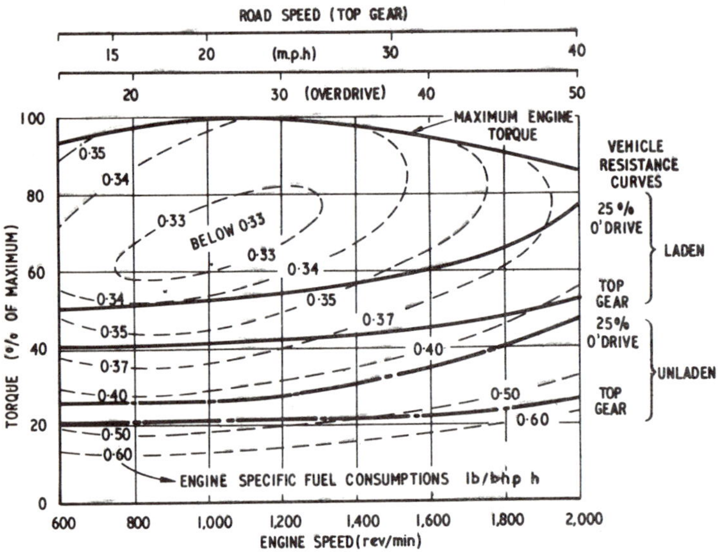

Fig. 5.3 Effect of overdrive gear for laden and unladen vehicles in comparison against diesel engine specific fuel consumptions

in overdrive (approx. 25% reduction in fuel). For the laden condition the figures are 0·41 and 0·36 respectively (12% reduction), but in both cases the vehicle speed is greater in overdrive gear.

5.4 DESIGN CONSIDERATIONS

Having established that vehicles may save fuel by employing a high gear for cruising conditions, there are a variety of alternative methods of achieving this high gear. It has been shown that in passenger cars the gains are sufficient to justify an additional gearbox, whereas in commercial vehicles it is more usual to provide an extra gear ratio in the gearbox or axle. Having decided to make provision for an overdrive, one of the first considerations is the ratio or size of 'step' to be introduced.

5.4.1 RATIO OF OVERDRIVE

It was seen in Fig. 5.1 that a 30% step-up can usually be accommodated without reducing the maximum speed of the vehicle on level ground and this appears to be one of the decisive factors. At road speeds less than maximum it will be clear that bigger step-up ratios than 30% could be employed with correspondingly larger gains in fuel economy and other benefits. The Borg-Warner overdrive (see Plate 5.2) utilises a step-up ratio of 42% and this overdrive has been widely used. Too high a step-up ratio reduces the power reserve for hills, acceleration or even headwinds so that more frequent changing back into direct drive is needed.

5.4.2 UNDERDRIVE VERSUS OVERDRIVE

Top gear in road vehicles is almost universally a direct drive so that losses due to gearing are not involved, and a high transmission efficiency is obtainable. By introducing an extra gearbox to step-up the transmission drive shaft speed it is necessary to introduce a gear train with some attendant gear losses. Some of the gain in fuel economy is lost in this gearing, and it would be logical to employ the most efficient transmission condition for economical cruising condition.

As a result, it has been suggested that an underdrive should be employed, and actual examples have been produced. Normal top gear operation could be obtained by a 30% reduction ratio in the transmission, and 'overdrive' operation could be achieved by locking this ratio into direct drive for best efficiency. To do this, however, it is necessary to make an appropriate 30% change in the axle reduction. From the vehicle manufacturer's point of view this technique requires the development of new axle reduction gears and further disorganisation of vehicle production by requiring a different axle when an overdrive is fitted or removed. It is possible to find examples of production vehicles which have different axle ratios when an overdrive is fitted as extra equipment, but these are small ratio changes and the vehicle performance would not suffer greatly if the overdrive was subsequently removed.

Until overdrives are standard equipment they are likely to remain overdrives, and not become underdrives.

5.4.3 POSITION OF OVERDRIVE GEARBOX

Overdrive gearboxes are usually placed on the output end of a vehicle's main gearbox where it may conveniently replace the long gearbox extension casing which is frequently used as a means of shortening propeller shafts. Situated in this position, it is usually necessary to ensure that the overdrive is never engaged when lower gears are in use so as to prevent excessively high torque being transmitted to the overdrive. Some engineers will consider that it would be more suitable for the overdrive to be inserted between the engine and gearbox, and some extremely small 'sandwich' overdrive gearboxes have been designed for this purpose. Fitting and removal of a sandwich overdrive is less easy because it involves more work in removing the main gearbox. Furthermore, the additional rotary inertia of overdrive gearboxes etc. on the input shaft imposes an additional duty on the gear-change synchronisers. The rearward position appears to be most convenient and there is not much disadvantage in restricting the use of the overdrive to the higher gears.

In certain vehicles, the overdrive has been moved further back to the rear axle where its weight helps to improve the front/rear weight distribution of the vehicle. The tendency to lower the height of floors in vehicles and the need to accommodate three persons in the front seat are better satisfied by the rear axle position. Transverse-engined vehicles have serious space problems and overdrives have not yet been adopted for this layout.

5.5 LAYCOCK OVERDRIVE

Overdrives manufactured by Laycock have been applied to a wide variety of vehicles and various different designs have been produced, but Fig. 5.4a shows a typical unit. A single epicyclic gear set is employed to obtain the step-up ratio and is arranged so that the planet carrier is connected to input, whilst an annulus gear is connected to the overdrive output shaft which drives the normal propeller shaft. A sun gear is employed to make the gearchange and is either held stationary for overdrive, or is locked to the annulus for the direct drive.

A compact arrangement is obtained by use of a friction cone

clutch attached to the control sun gear. This cone slides along splines on the sungear shaft and engages either with a brake ring mounted on the outercasing, or at the other extremity of its movement, with a cone on the annulus gear. The friction cone thus works internally and externally and resin-bonded friction material

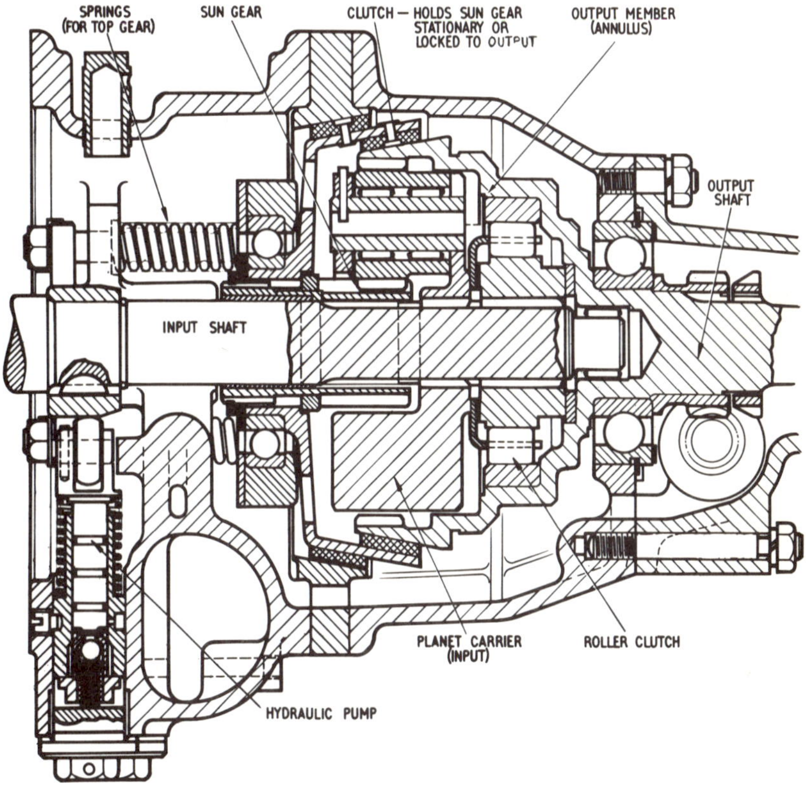

Fig. 5.4a Sectional illustration of the basic components in the Laycock overdrive gear

is riveted to both faces of the cone. This cone is loaded by springs towards the annulus cone (direct drive) but a hydraulic piston is used to move the cone against these springs towards the brake ring for overdrive operation. In certain versions, such as shown in Fig. 5.4b when helical gear teeth are used, a resultant thrust acting on the sun gear is employed to act against the return springs. As this thrust reduces progressively when the sun gear accelerates up to annulus speed, the spring force increases in a complementary

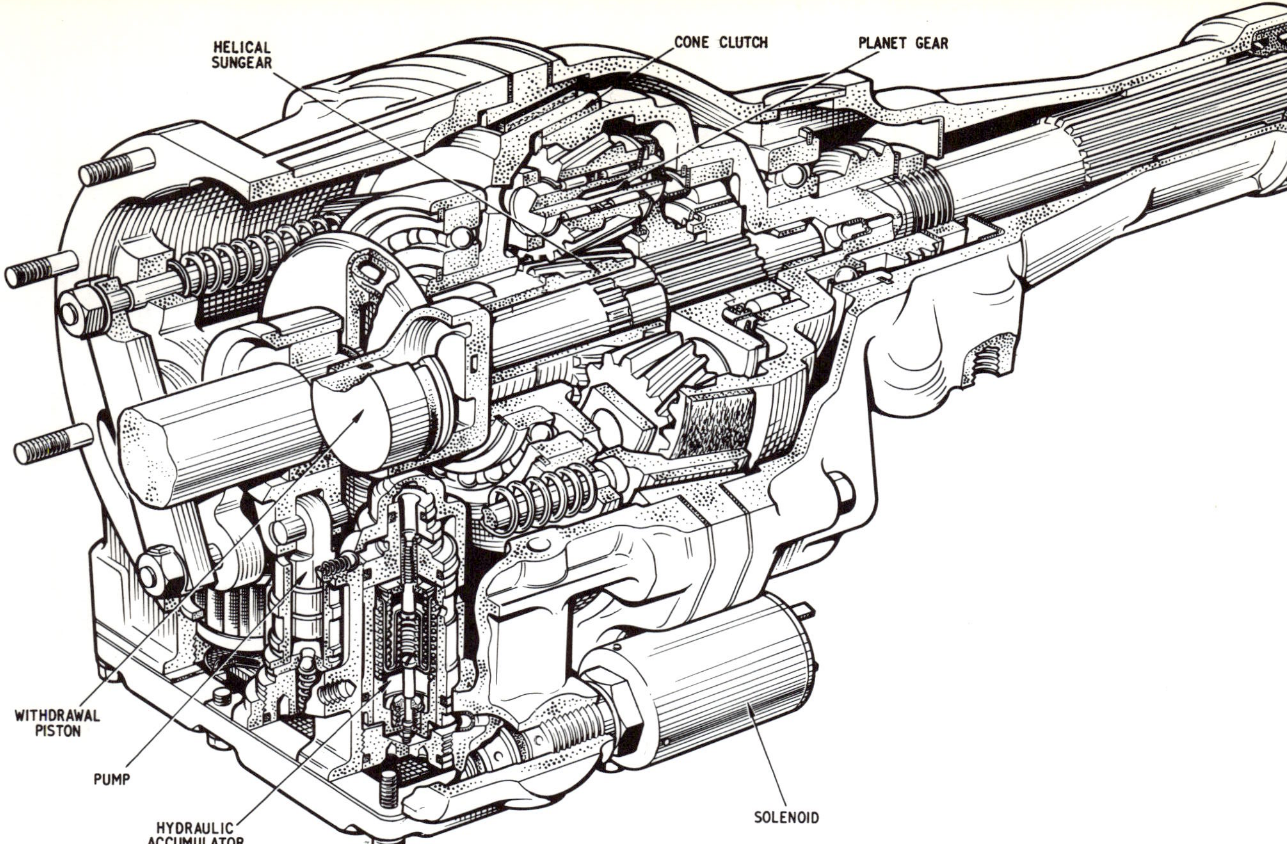

Fig. 5.4b Three-dimensional view of a complete overdrive gearbox. (Laycock)

fashion so that maximum load is reached after the gearchange is complete.

A roller clutch is mounted inside the annulus member and ensures that the input shaft cannot rotate faster than the output shaft of the overdrive. As well as a precaution against clutch slip, this device maintains torque during an upchange until the friction cone is brought to rest in the brake ring.

5.5.1 CONTROL

Hydraulic pressure is obtained from a small single-plunger pump which charges a hydraulic accumulator and thus stores sufficient fluid at pressure to operate two hydraulic pistons and move the friction cone on to its brake ring.

Overdrive gear is introduced by an electric solenoid which, when energised, moves a lever to lift an operating valve and ball to admit

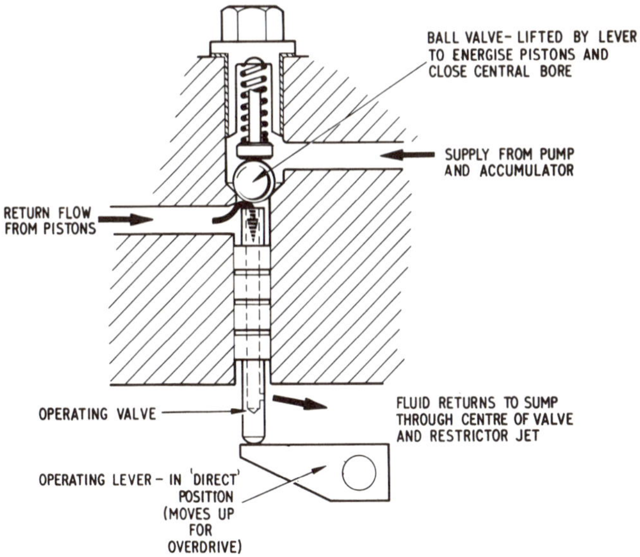

Fig. 5.5 Sectional view of an overdrive operating valve. (Laycock)

pressure to the operating pistons. As seen in Fig. 5.5 the operating valve is drilled with a concentric bore and restriction jet so that when de-energised, the operating lever falls, leaving the ball on its seat to prevent flow of fluid under pressure from the pump. Fluid returning from the pistons under the action of the clutch springs returns to the sump reservoir through this central bore and restriction jet. The size of the jet is chosen so that the clutch takes about 0·5 sec to return to drive from the overdrive position.

In some Laycock units, see Fig. 5.6, the hydraulic accumulator acts as a relief valve and when the accumulator piston is deflected fully against its spring, a series of holes or ports in the accumulator cylinder are disclosed providing a flow of fluid for lubrication or cooling etc. More recent designs have added further improvements to the accumulators, which can act as a reserve supply of fluid to move the operating pistons quickly into position as well as acting as a smoothing device.

5.5.2 SEMI-AUTOMATIC OPERATION

In the Laycock overdrive, the gearchanges are not wholly 'hot-shift' since although torque is maintained during the upchange, the drive is disconnected during the downchange, whilst the clutch moves from one cone to another. The provision of the freewheel

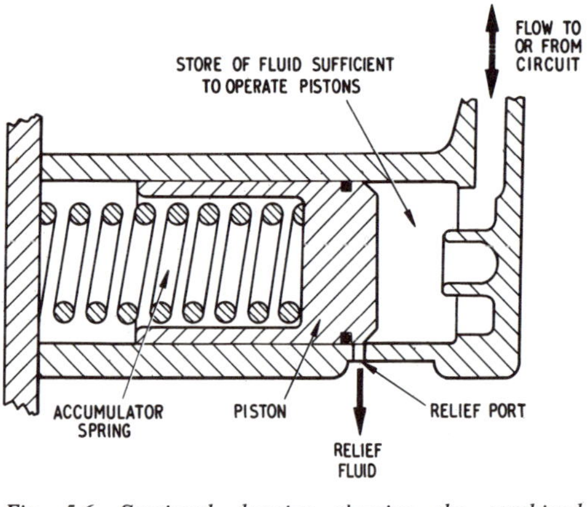

Fig. 5.6 Sectional drawing showing the combined hydraulic accumulator/relief valve. (Laycock)

unit enables engine torque to be transmitted during the upshift and thus qualifies for the semi-automatic title. Fully-automatic operation of a gearchange requires two separate friction or coupling elements, one of these elements being disengaged progressively whilst another is smoothly engaged.

The Laycock unit having only one friction member cannot

achieve this double action, but it approaches very closely the operation required for a fully automatic gearchange.

5.6 BORG-WARNER OVERDRIVE

The general layout of the Borg-Warner overdrive is similar, and an 'exploded' view shown in Plate 5.2 shows some similarities

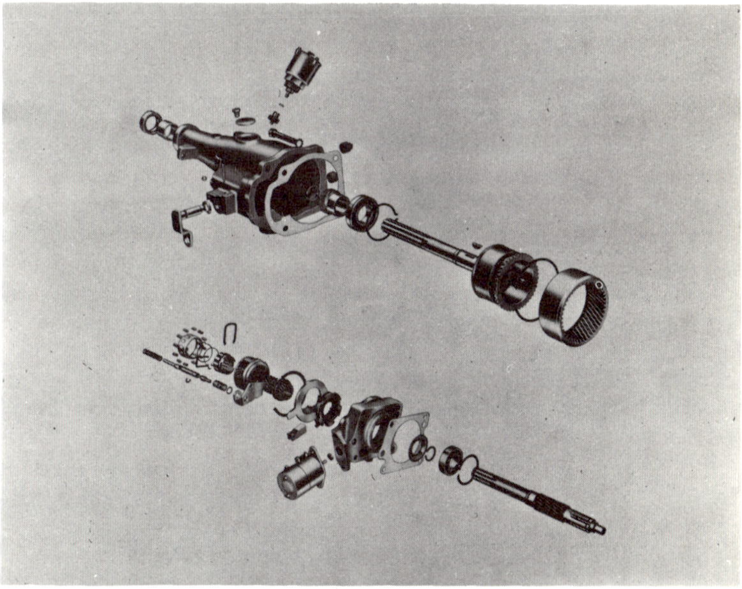

Plate 5.2 Exploded view of the Borg-Warner overdrive gear. (By courtesy of Borg-Warner Ltd.)

in position and arrangement to the Laycock designs. It employs a simple sun, planet and annulus gears arranged so that the sun gear acts as a reaction member and the annulus and planet carrier act as output and input members respectively. A 'sprag' type of one-way clutch (freewheel) is fitted between output and input members so that the input shaft cannot overrun the overdrive output shaft.

Gearchanges in the Borg-Warner overdrive are of the power-off type, and the driver lifts his foot from the accelerator pedal to engage overdrive when travelling above a certain predetermined road speed. For a downchange into direct drive, the accelerator is fully depressed, but the engine ignition is cut momentarily to release the torque on the sun gear reaction member, which is then unlocked.

Direct drive is achieved by the overrunning clutch, and the sun gear reaction member is not normally locked to any other part for normal top gear running. In this condition the sun gear can spin freely and the overdrive 'freewheels' on down grades or if the engine speed is reduced. A separate mechanical control is provided to slide the sun gear into toothed mechanical engagement with the planet carrier in order to provide a locked up position for the overdrive for reversing or other special operation. This control cannot be operated when the overdrive ratio is in operation.

5.6.1 CONTROL

As the gearchange is operated in a power-off condition, a simple mechanical engagement or release of the sungear is effected by means of an electrical solenoid. When energised, this loads a spring on a single mechanical pawl which engages with a toothed 'dogplate' in the sungear as soon as this gear stops rotating as a result of lifting the accelerator pedal. If the electrical solenoid circuit is broken the pawl is retracted and the sungear is released, aided by a momentary 'engine cut'. A small centrifugal switch breaks this circuit when the vehicle speed drops below about 30 m.p.h. There are thus four separate control items employed in the control of this overdrive:

(1) Centrifugal switch	closes switch in solenoid circuit after preset speed is attained (28–35 m.p.h.)
(2) Kickdown switch (1)	breaks solenoid circuit.
(3) Kickdown switch (2)	breaks ignition circuit momentarily.
(4) Panel control knob	slides sungear into mechanical lock with planet carrier for direct drive.

In addition, the operation of the overdrive requires a release of torque in the sun gear for a gearchange and this release of torque is achieved by:

(1) lifting foot from accelerator pedal for upchange.
(2) fully depressing accelerator pedal to operate the kickdown ignition switch for downchange.

5.7 OVERDRIVES FOR COMMERCIAL VEHICLES

As mentioned previously, it is more usual to make provision for overdrive operation by providing additional ratios within the gearbox and it is not infrequent to find a step-up ratio in such gearboxes of 1·4 or 1·5 overdrive. In some 10-speed gearboxes two overdrive ratios may be included, but the maximum overdrive ratio does not usually exceed 1·50, and the additional gears are employed to reduce the steps within the available range of ratios.

Some commercial vehicles employ a two-speed axle which provides two separate reduction ratios (see Chapter 7). The additional ratio is obtained usually by means of an epicyclic gear set employed in a very similar way to those used in the previously described overdrives. A change in the axle ratio is made under power-off conditions by a sungear reaction member which is held stationary or driven at input speed by mechanical tooth engagements. The epicyclic gear set is combined with the helical bevel or hypoid gearing in the axle casing and an electrical solenoid or vacuum pressure is employed to make the gear change. Such axles will be described more fully in a later chapter but the gear step thus obtained is equivalent to a 35–40% overdrive and is popularly used for fast, unladen, return journeys. Three speed axles are also available but these are less well known.

Despite the above, it is still possible to find overdrive units designed as a separate unit and for commercial vehicle use. One such example is seen in Fig. 5.7 which shows an automatic or 'hot shift' gearbox combined with a fluid coupling to provide a direct drive and a 40% overdrive ratio.

This overdrive can be engaged or disengaged under full power conditions since it has two friction elements, one a multiplate clutch and the other a friction band brake. With these two elements it is possible to arrange for a transition from one ratio to another whilst maintaining a power output. Semi-automatic operation is normally employed using a small switch or lever operated by the driver to select 'direct or 'overdrive' as required. It will be clear from the illustrations that the overdrive is mounted behind the engine and operates 'in front' of a gearbox or other transmission required by the vehicle.

5.8 OPTIMUM ECONOMY

It can be concluded from the foregoing that the various types of overdrives and methods of obtaining a similar effect take advantage of the fact that vehicles frequently cruise at less than maximum speed and that a better fuel economy can be obtained from an engine by use of higher gears. As seen from earlier graphs, throttling an engine is an inefficient method of control and may halve the

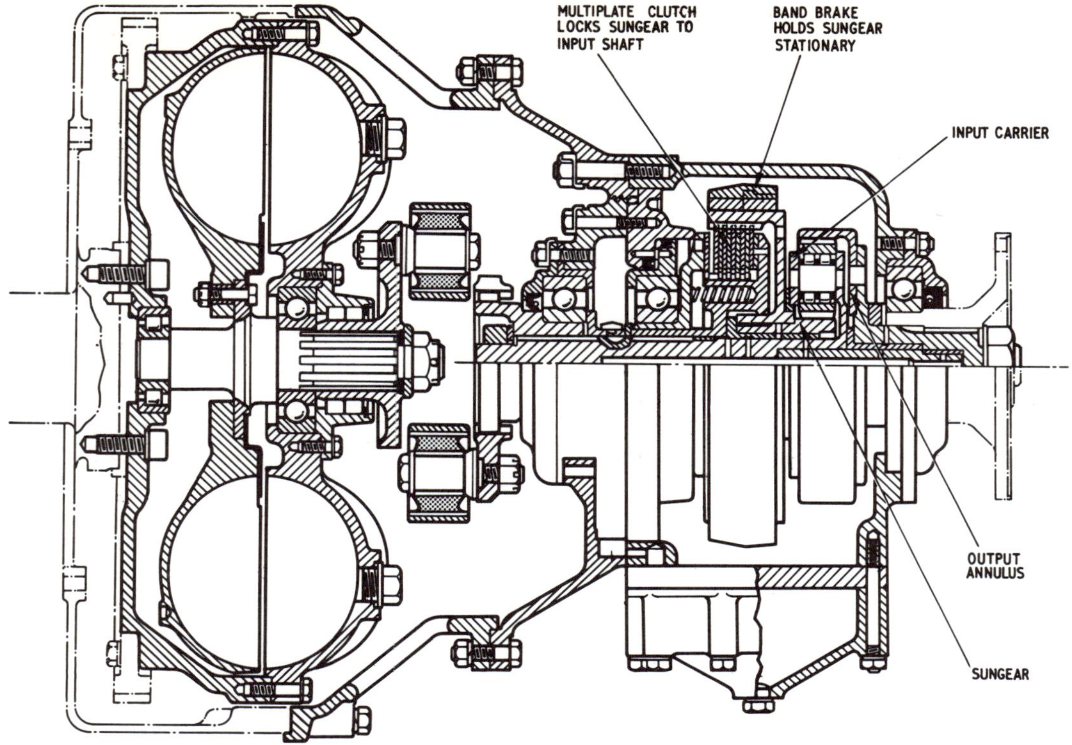

Fig. 5.7 Part-sectional view of an overdrive gear suitable for commercial vehicles. This includes a fluid coupling and gear changes are fully automatic and without interruption of power. (Leyland Motors Ltd.)

efficiency of a diesel engine and have a worse effect on a petrol-engined vehicle.

In endeavouring to obtain improvements in operating economy by introducing an extra gear ratio it has been noted that the choice of this ratio, or overdrive, is made as a compromise between what is needed for a power requirement curve of a typical journey and balanced against considerations of the proportion of the journey for which the overdrive can be engaged and how many gear changes are needed to obtain the best operation. Whereas a 50% overdrive for commercial vehicles and 35% for passenger cars seems to be a practical limit, it is easy to show that much higher overdrive ratios could be employed. From Fig. 5.2 it will be noted that a vehicle at 76 m.p.h. requires about 26 bhp to maintain this speed on a level road and with normal gearing obtains this power at a specific consumption of 0·60 lb/bhp h. This engine power could be achieved at a specific consumption of only 0·48 lb/bhp h but since the engine speed must be reduced to about 2,000 rev/min instead of 3,800 rev/min an overdrive ratio of 90% would be needed. Also from this figure since the power requirement of the vehicle is under 13 hp until 57 m.p.h. is reached on the level, there is little need from an engine operation point of view in exceeding 1,000 engine rev/min which is the speed at which it delivers this power. To achieve this type of operation overdrives of nearly 200% could be utilised. The gains in fuel economy obtained by these extremely high overdrives would scarcely justify overdrive ratios of 2 : 1 or 3 : 1 speed increase, because the time and fuel spent at these conditions is small. Greater gains in operating economy would be obtained from some form of variable ratio drive which would permit an engine to operate always within its most economical regions.

In the case of an overdrive 'step' ratio, a driver must always be considering whether he should engage 'overdrive' or 'direct' for each part of his journey and he would need to make frequent changes between these two gear ratios. If a variable ratio overdrive was available, then this device could adjust itself to variations in gradient or to suit the driver's needs. Clearly, there would be little point in extending the overdrive ratio beyond about 50% speed increase but most operating conditions and open road travel could be satisfied by a transmission ratio variable within the range between direct drive and 50% overdrive.

5.8.1 OPTIMUM ENGINE UTILISATION

The ultimate in engine utilisation is to operate an engine at the most economical setting for each power setting. Thus it is possible to draw an optimum utilisation or maximum efficiency line on the torque or power curves of an engine to select all points of maximum engine efficiency within the engine power range.

Fig. 5.8 shows how this maximum efficiency curve may be compared with the normal engine power curve. It will be noted that the maximum efficiency curve meets the power curve at the point of maximum engine output power, and hence there is no loss in engine or vehicle performance. It will be noted, however, that the maximum efficiency curve is achieved at less than full throttle setting apart from this one point of coincidence at maximum output.

Assuming an axle ratio such that the maximum power and maximum vehicle speed are achieved at the usual direct drive then Table 5.3 shows the correct overdrive ratio needed to maintain the engine on its maximum efficiency curve.

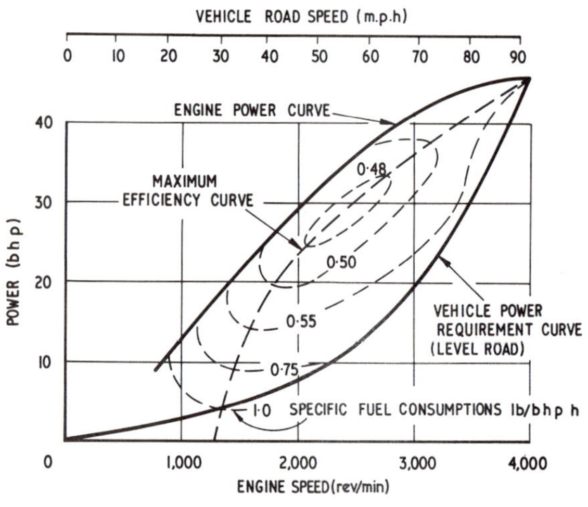

Fig. 5.8 A 'maximum efficiency' curve for engine utilisation. (Optimum engine utilisation curve.)

Table 5.3 shows that a 60% overdrive is sufficient to achieve maximum economy on this particular vehicle and it seems probable than a 50% speed increase would be sufficient to obtain satisfactory gains.

Below 30 m.p.h. a satisfactory engine efficiency is attained at higher engine speeds so that the overdrive ratio is not needed.

There is no variable ratio overdrive generally available so that the Table 5.3 can only be used as a guide in the choice of a suitable fixed overdrive gear ratio. As will be seen in the following chapter there are various attempts being made to develop variable ratio transmissions which include 'overdrive' or speed-increasing ratios and Fig. 5.8 and Table 5.3 illustrate the basic requirement for the automatic control of variable ratio transmission.

An automatic or variable transmission would endeavour to follow the optimum utilisation curve for all road conditions. By

Table 5.3 OVERDRIVE REQUIREMENTS FOR MAXIMUM ECONOMY

Road speed (m.p.h.)	*Propeller shaft speed* (rev/mm)	*Vehicle power Requirement* (bhp)	*Engine speed for Maximum Efficiency* (rev/min)	*Overdrive ratio needed*
20	870	2·2	1,300	0·67
30	1,310	4·0	1,350	0·96
40	1,740	6·3	1,400	1·24
50	2,180	9·4	1,450	1·50
60	2,610	14·0	1,600	1·63
70	3,050	21	1,900	1·61
80	3,480	30	2,500	1·39
90	3,920	43	3,650	1·24
92	4,000	46	4,000	1·00

considering the maximum hill climbing power requirement of the vehicle at various road speeds the maximum torque multiplication ratio of a variable ratio could be determined which, in conjunction with the overdrive ratio table, would fix the total range of variation of the transmission.

5.8.2 AUTOMATIC SELECTION OF MAXIMUM EFFICIENCY

Before leaving the subject of engine efficiency it is necessary to discuss the combinations of engine speed and torque which give use to maximum efficiency. Engines vary, of course, in their operating economy and alternative methods of mixing or metering fuel will have an effect. The diagram in Fig. 5.8 however is typical of many carburetted engines and maximum efficiency is obtained by a fairly rapid increase of torque from zero to 0·75 maximum between 1,000–2,000 rev/min, followed by a steady increase in throttle opening with engine speed until the maximum power condition is obtained.

The governor or controller of automatic or variable ratio transmissions should be set to adjust the operating transmission ratio so that at any road speed the engine will produce the required power at the correct speed combination. There are many ways of obtaining the required combinations but all types of governor or selector must be provided with information about engine speed and engine torque.

Table 5.4 shows the optimum condition obtained from an engine which delivered a maximum power of 56 bhp at 4,200 rev/min.

Table 5.4 ENGINE CONDITIONS FOR MAXIMUM ECONOMY

Power demand (bhp)	*Engine speed* (rev/min)	*Engine torque* (lb ft)	*Specific fuel consumption* (lb/bhp h)
5	1,075	24·5	0·85
10	1,225	43	0·62
15	1,400	56·0	0·54
20	1,650	63·5	0·515
30	1,260	73	0·474
40	2,735	76·5	0·53
50	3,405	77	0·57
56	4,200	70	0·62

Centrifugal governors are frequently used to obtain a measure of engine speed, but measurement of engine torque is less easy. The general convex shape of most engine torque curves makes them unsuitable for control on grounds of stability. Throttle opening or inlet manifold vacuum pressure can be employed to obtain the required combinations.

This chapter has devoted space to the study of operating economy as well as in discussion of the overdrive gearboxes. The economy and other benefits obtained with the overdrive unit tend to be reversed by some automatic transmissions which usually show slight increase in fuel consumption. Hence there has been a tendency for vehicle manufacturers to offer automatic transmission in overdrive as alternative extras, since they apparently serve opposing interests. The last paragraphs have attempted to show that automatic control of transmission ratio (including overdrive ratio) can be chosen to give good operating economy. As progress moves towards variable ratio transmissions then the opposition between overdrives and automatic transmissions may disappear. The good fuel economy of automatic transmission in traffic could be combined with the high speed economy of overdrive to obtain a much superior transmission performance.

The widespread use of overdrive gears is a reminder to automobile engineers that reduced noise and improved economy are appreciated by customers. The better operating economy is an excellent lesson in engine operation. Thus although engineers have spent much time in getting better engine performance and satisfactory fuel consumption, their best thermal efficiencies are still only around 30%. Parts of the engine's performance fall below 10% thermal efficiency and although these low efficiencies do not form a large part of a vehicle's operating range, it is fairly clear that engines do operate in regions of lower efficiency for long periods.

Overdrive gears show an operating gain by pushing the engine speed and torque into the more economical areas. Transmission designers should perhaps give more consideration to these ideas in choosing greater ratios for a given vehicle and even perhaps to transmissions which can maintain the engine in its more efficient region. A variable transmission having a 10% loss is a worthwhile consideration if it can give an improvement to the engine efficiency from 20% to 30%. Thus although the transmission would reduce this 30% to 27% thermal efficiency, the overall gain is from 20% to 27%—a 35% increase in economy.

5.9 LOAD FACTOR

This chapter also emphasises that road vehicles operate at a very low proportion of the maximum engine output power. It is easy to deduce the overall load factor of a vehicle from its fuel consumption in gallons per hour. If a vehicle covers a journey at 30 m.p.h. average speed and 25 m.p.g. fuel consumption then it is burning fuel at 1·2 gal/h, i.e. about 8·4 lb/h. Now the engine performance graphs showed that 0·6 lb/bhp h is a reasonable mean figure for the specific fuel consumption. Hence the engine was working at an average rate of $8{\cdot}4/0{\cdot}6 = 14$ bhp. This is a very low proportion of the engine's maximum power which in most cases is likely to be 50 bhp or more.

By assuming that petrol engines produce 12 bhp h from a gallon of fuel and that diesel engines can produce 20 bhp h, then Table 5.5 shows that most internal combustion engine vehicles operate at very low load.

Obviously the load factor is a mixture of high rates of working and lower rates of working, but it is clear that engine economy at low load factors is an important feature of vehicle operating economy.

Table 5.5 TYPICAL LOAD FACTORS

Vehicle	*Maximum engine power* (bhp)	*Typical route speed* (m.p.h.)	*Average fuel consumption* (m.p.g.)	*Fuel burnt per hour* (gal)	*Mean output power* (bhp)	*Load factor* %
Racing car	250	80–100	10	8–10	96–120	40–50
Off-highway dumper	120	—	—	3	60	50
Industrial truck	40	—	—	0·5	10	25
Heavy truck	120	25	8–11	2·3–3	46–60	40–50
Omnibus	100	12	10	1·2	24	24
Light van	50	30	25	1·2	14	28
Large car	120	50	20	2·5	30	25
Small car	50	40	28	1·43	17	34
Motor cycle	30	40	60	0·67	8	27
Moped	5	15	120	0·12	1·4	28

BIBLIOGRAPHY

1. ASHBY, S. H., 'Overdrives—Problems in their Application and Design', *Proc. Inst. Mech. Engrs. (London) Automob. Div.* (1956)
2. ANON., 'Birfield Group Overdrives', *Automob. Engr.*, **57,** No. 11, 419–20 (Oct., 1967)
3. ANON., 'An Overdrive from America', *Motor,* 285, 28 Sept. (1955)
4. MERRELL, R. L. and BOWRIE, E. F., 'Roller One-way Clutches for Today's Passenger Car Automatic Transmissions', *Auto. Engr. Congress,* 9–13 Jan. (1967), *S.A.E. Paper,* No. 670049.
5. ANON., 'Underdrive Splitter Gear—Thorneycroft,' *Automob. Engr.*, **56,** No. 9, 359 (Oct., 1966)
6. GILES, J. G., 'Transmission Ratio Requirements of Road Vehicles,' *MIRA report* (1957/4)
7. GILES, J. G., 'Variable Ratio Transmissions and Vehicle Operating Economy', *Mobil Research Symposium, Amsterdam,* 3–4 May (1966)

6

variable ratio drives and fluid transmissions

Automobile engineers have long known that the ideal vehicle transmission would be one whose operating ratio could be adjusted to any value within a required range. Many differing mechanisms have been tried but it is only recently that some of the possible designs have shown themselves capable of competing with current transmissions on grounds of size, weight, cost and reliability.

Although mechanical gearboxes have been the typical transmission for road vehicles for a long time, the search for a variable ratio transmission is becoming more urgent. Mechanical gearboxes are growing more and more complicated by the need to include more gear ratios, or to achieve torque-maintained (hotshift) and automatic gearchanging. Most types of vehicles show a need for a variable ratio transmission and usually because of a road performance requirement. Some differing vehicle applications are shown below together with some of the reasons why a variable ratio drive is desirable.

Passenger cars. Desire for effortless driving—especially transmission controls —without loss of performance or economy.

Passenger service vehicles. Smoother ratio changes, low maintenance, easier driving, improved economy in dense traffic.

Commercial vehicles. Need for full engine power over wider range of load/speed combinations without large gaps or steps between ratios. Torque maintained during ratio changes (hot-shift).

Tractors and off-highway vehicles. Torque maintained gearchanges or ratio variation. Automatic and rapid ratio variation to match changes in load, terrain etc.

Racing cars. Full engine power required over increasing range of roadspeeds with a decreasing range of usable engine speeds (hot-shift).

In addition to the above, any vehicle, including scooters and cycles would be well served by a cheap and simple variable drive with automatic gears, and easy starting properties. Although experimental or pilot production transmissions have been employed for all the previous list of vehicle applications, only a few examples of a truly variable drive have reached large scale production. Amongst the wide variety of variable ratio drives which have been considered, a representative sample will be discussed and described under the following headings which appear to cover most of the important designs.

(1) Friction and oilfilm drives
(2) Hydrokinetic fluid drives
(3) Hydrostatic fluid drives
(4) Hydromechanical variable drives

6.1 FRICTION AND OILFILM DRIVES

Friction drives are usually of the variable pulley type in which the sides of a vee-pulley may be moved to adjust the width, and hence the effective diameter, of a belt pulley wheel in order to adjust the speed ratio of a pair of such pulleys. Oil film drives comprise of a number of wheels or rollers which are loaded into contact with a driving wheel so that the path of contact may be adjusted to different radii of the driving wheel. Although this type is still occasionally referred to as a friction drive, experience shows that there is a small relative movement between the contact faces of the driving wheel and of the driven wheel or roller and that the output torque of the drive is achieved by the shearing of a very thin oil film which separates the two faces.

6.1.1 DAF VARIOMATIC TRANSMISSION

As seen in Fig. 6.1 the DAF Variomatic transmission, which is part of the interesting Dutch car of that name, operates on the well-known variable pulley system. The width of the two pulleys may be adjusted so that the vee-belt can operate at different effective diameters. The driving pulley (top) is adjusted by centrifugal bob-weights which tend to move outwards and increase the operating radius of the driving pulley when its speed increases. The driven pulley is spring loaded to operate at maximum diameter. The operating diameter of the driven pulley is thus determined mainly by the position of the driving pulley and the springs ensure that there is sufficient tension in the vee-belt, and resist the outward movement of the bobweights.

Thus, the effective diameter of the pulleys is controlled by the centrifugal weights on the driving pulley acting against this spring

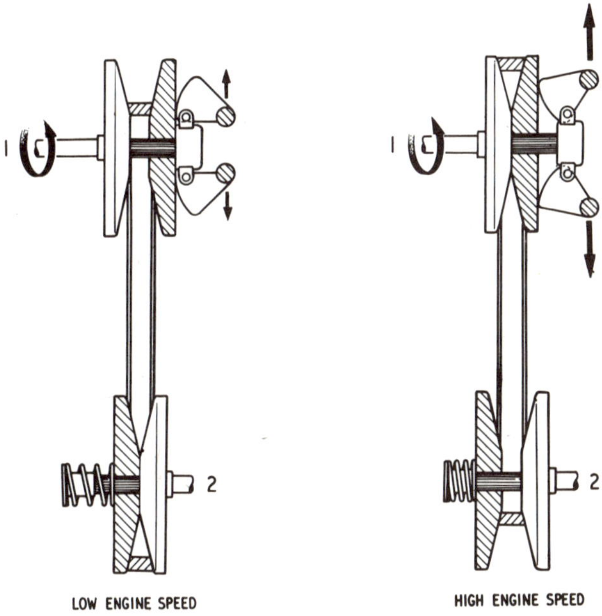

Fig. 6.1 Basic principles of the variable-pulley belt drive: (1) Driving pulley with centrifugal weights; (2) Driven pulley spring-loaded

on the driven pulley. Manifold vacuum pressure is also employed to oppose the centrifugal load so that the engine speed (and driving pulley speed) is higher when the accelerator pedal is depressed. In this way, the transmission ratio is automatically adjusted to obtain maximum engine power whenever the accelerator pedal is fully depressed (low vacuum pressure). At lower throttle openings, the engine speed is adjusted to produce engine power at economical settings in a manner similar to that described in the previous chapter.

As shown in Fig. 6.2, the full engine speed (hence power) is available at most road speeds by depressing the pedal fully. For steady cruising, the engine speed is much lower and does not often rise much above 2,000 rev/min until the road speed approaches the vehicle's maximum speed. At any accelerator pedal position, the vehicle settles down to a steady speed on a given road or gradient. Further downward movement of this pedal will cause the vehicle to accelerate. The amount of acceleration will be dependent on the amount of further downward movement of the accelerator pedal. The accelerator pedal is thus correctly termed in this vehicle and movement up or down for any steady condition will decelerate or accelerate the vehicle in the normal sense. Except for very low road speeds, full depression of the pedal represents full engine power. It will be noted from Fig. 6.2 that the use of

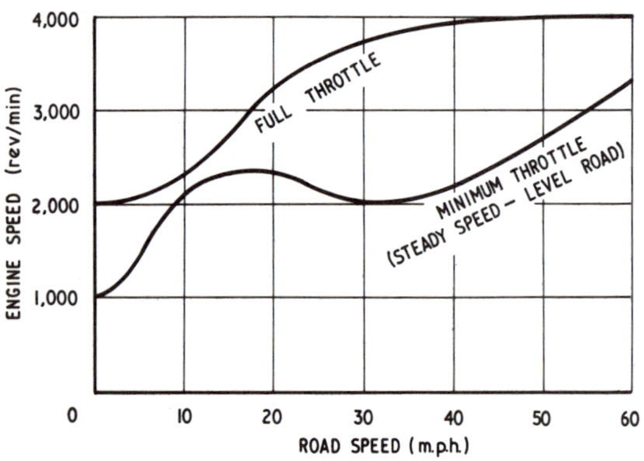

Fig. 6.2 Control characteristics of DAF Variomatic transmission. Engine speed for maximum/minimum conditions at each road speed

engine vacuum for control against a centrifugal device does not conform exactly to the optimum requirement and engine speed is rather high in the 10–20 m.p.h. road speed range. However, the control mechanism gives a very close and satisfactory approximation to this requirement and otherwise closely approaches the ideal variable ratio transmission enabling full engine power to be available when required and at other times using the engine in its economical range. An efficiency curve for the transmission is shown later (see Fig. 6.5) but is mainly above 90% and is satisfactory in this respect.

The use of engine manifold vacuum pressure tends to give a rather 'throttle sensitive' control and a recent improvement included an additional pressure reservoir which reduced the number of small adjustments of engine speed which can occur under certain conditions. Vacuum pressure is generated when the engine is used for braking on long descents, etc., and this pressure can be used to speed up the engine to assist the vehicle's brakes. Stored vacuum pressure in this reservoir may be applied to 'down-change' the transmission by operating a facia board lever or by depressing the brake pedal. In this way, the engine speed increases to give more retardation to the vehicle.

The DAF vehicles are comparatively low powered (below 30 bhp) and do not have high maximum road speeds. Reports of many users, and in journals, comment on the excellent acceleration for traffic use when these vehicles show no loss in performance in comparison with much more powerful vehicles. In view of the fact that the 750 cm^3 engine vehicle has a top speed of not much over 60 m.p.h., an acceleration time 0–30 m.p.h. in 4 sec is quite remarkable. The variable ratio transmission helps in achieving this quick acceleration because there are no pauses in the use of the engine which continues to work at near maximum power whilst the transmission ratio is being adjusted continuously to convert this power into thrust at the road wheels. Table 6.1 shows the transmission ratios obtained from the vehicle at full throttle and minimum throttle settings with axle gearing chosen so that 4,000 rev/min (engine) in top gear (1 : 1) was equivalent to 55 m.p.h.

Table 6.1

Road speed (m.p.h.)	*Transmission ratio (Reduction)*	
	Full throttle	*Minimum throttle*
5	5·75	3·8
10	3·15	2·8
15	2·55	2·1
20	2·35	1·58
25	1·92	1·18
30	1·70	0·92 (9% O'drive)
40	1·34	0·72 (38% O'drive)
50	1·10	0·75 (33% O'drive)
55	1·00	0·75 (33% O'drive)

It will be seen from this table that the Variomatic transmission provides a good range of transmission ratios from a low gear varying through to overdrive. Fig. 6.3 shows a sectional drawing of the transmission which is combined with the rear axle assembly and suspension. The variable pulleys and belts are situated between two gear reductions. The drive from engine is taken into a bevel

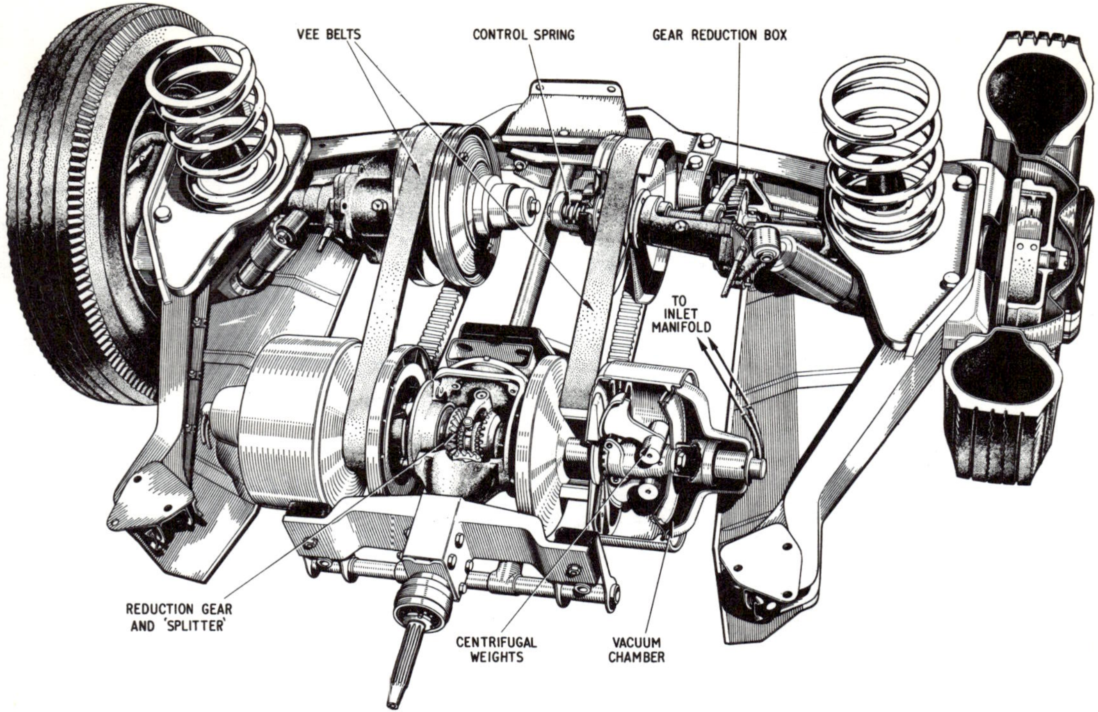

Fig. 6.3 Sectional illustration of the Variomatic transmission (DAF)

reduction gearbox and drives two variable pulleys. By sliding the bevel gears along a shaft, a forward, neutral, or reverse selection is obtainable. Two V-belts are used—one to each rear wheel and a reduction gearbox connected to each output pulley may rotate about the pulley axis to act as the swing arm of a rear suspension member. Situated between these two gear reductions, the variable pulley transmission operates over a favourable range of reduction and speed-up ratios. Belt torques are minimised by employing roughly equal amount of variation above and below the direct drive (1 : 1) condition.

An adjustment is provided to take up any stretch in belts or to allow for variations in manufacture of belt length. The spring-loaded output pulleys maintain a constant tension in the belt and also have a differential action when cornering to permit one wheel to rotate faster than another.

Although there are still doubts about the life of belts when employed for higher engine powers, it is quite obvious that 10–15 bhp per belt is being successfully transmitted in these DAF vehicles. Their performance does not compare with most other small cars but the experimental use of the Variomatic transmission in a 1 litre racing car is an expression of confidence and may do much for this type of transmission.

6.1.2 PERBURY GEAR

This oil film drive is illustrated in Fig. 6.4 from which it will be seen that an input shaft brings a drive from the engine to a central driving disc which is machined to a toroidal surface on front and rear faces. This toroidal shape is a circular arc when viewed in a cross-section across a diameter of the disc. This curved surface drives six rollers mounted on a carrier and in contact with two toroidal end discs. The whole assembly is loaded together with a total load of several tons which is sufficient to provide an output drive on the two end discs which are mounted on the central mainshaft. The rollers are arranged in two sets of three and each roller may be rotated so that it makes contact with any part of the toroidal surfaces.

Both sets of three rollers are connected so that all rollers are making contact with the toroidal discs at the same circumferential path on each disc. In this way the drive is shared between the rollers and the torque multiplication obtained is proportional to the ratio of the circular paths of contact on the output/input discs. As in the case of the variable pulley drive it is convenient to provide an

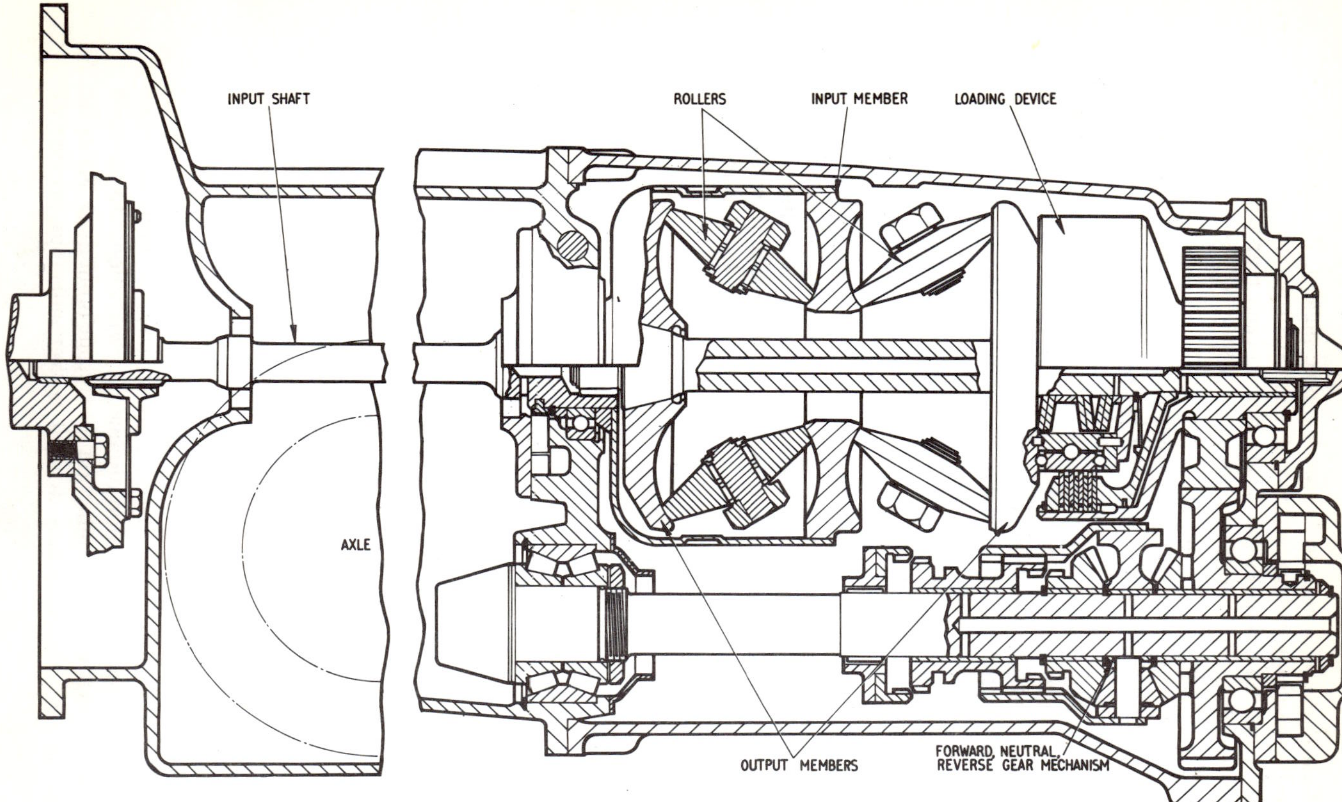

Fig. 6.4 A typical Perbury oil-film transmission

equal amount of variation above or below the direct drive setting which is obtained with the rollers set in a horizontal position.

The rollers are mounted on a common carrier frame which receives a reaction torque equal to the combined input and output and is held stationary by the transmission casing which in turn must be fixed to the engine casing or the structure of the vehicle. The rollers take up a new position or ratio by a steering action. This steering action is obtained by moving the rollers so that their central bearing axis moves off centre from the longitudinal axis of the transmission, and is achieved by rotating a concentric hollow shaft inside the carrier frame. This hollow shaft is rotated a few degrees from a central position for the rollers to tilt themselves to a higher or lower transmission ratio.

The discs and rollers are made of the hardest steels which must be very accurately shaped and machined. It will be noted that the rollers have a width of about 0·25 in on the contact periphery. When in contact with a toroidal disc each roller is covering a number of possible contact paths. In practice the actual output ratio is equivalent to some mean peripheral path within that contact band or area, of the toroidal disc covered by the roller during one revolution of the disc. Since the circumference of a circular path around a toroidal disc is proportional to the radius of that path from the centre of the disc, then there must be 'slip' or relative movement between the roller and disc at both the inner and outer edges of the roller. The roller and disc are rotating at some relative speed fixed by the roller circumference and the circumference of a mean path on the disc. The inner edge of the roller is moving faster than the disc, and the outer edge of the roller is moving more slowly.

This relative motion (slip) occurs in various other rolling actions —such as taper roller bearings for instance. In practice it appears that there is slight 'slip' or relative motion across the whole face of the roller and torque is transmitted by the shearing of the fluid film existing between the metal surfaces. This fluid film is extremely thin and the mechanism of the transfer of torque in the contact area is a complex study but special fluids will produce higher output torques with the same axial load on the disc and rollers, and mathematical analysis can help in the mechanical design.

The amount of slip needed to generate torque is very small (about 2%) and the efficiency of the drive is high, although there are some losses developed in the oil film and bearings, etc. so that a typical efficiency curve is shown in Fig. 6.5. This curve shows the full load efficiency, ignoring auxiliary losses due to gears, lubrication, etc., and it seems that there is only a few per cent drop in efficiency at half load. The graph also gives an efficiency curve for the DAF Variomatic transmission. In all types of friction or oil film drive there must be some 'slip', because the contact between the working surfaces is made over an area. A V-belt develops torque by friction between the side of the belt and the pulley. Since a belt must have some thickness for strength, the top layers and bottom layers must be slipping to conform to the mean belt speed.

Both types of drive appear to have a limited life. The slip in a belt eventually causes the rubberised fabric to fail, usually due to loss of rubber. In the oil film drive the heavy axial loads lead to surface fatigue failures (pitting) after a given period of time, and the size of the transmission must be carefully chosen in order to obtain a satisfactory life.

A variable ratio drive in itself is not sufficient for an automobile transmission which, as mentioned before, requires additional features for starting, control of the transmission ratio and reverse gear etc. In the small DAF vehicle, simple centrifugal devices are employed for actuating a friction clutch (two stage) for starting, another centrifugal device, acting against inlet manifold pressure, selects and adjusts ratio whilst the bevel gear drive for the rear axle makes a simple provision for reverse gear. The Perbury gear needs the same additional features, which together with a hydraulic power supply and a need to prevent excessive overloads, add some complexity to the comparatively simple basic mechanism.

6.2 HYDROKINETIC FLUID DRIVES

This type of fluid drive has already been described briefly in Chapters 2 and 4 where some of the simpler designs were illustrated in connection with their use for the starting of vehicles or in connection with automatic transmissions. There have been many different variations of the basic hydrokinetic drive in addition to the basic coupling and converter designs already illustrated.

Hydrokinetic fluid drives consist in principle of a number of turbines or water wheels situated in the flow path of the circulating fluid. The initial flow of fluid is generated by an impeller or pump member which is a special version of a normal centrifugal pump impeller. The impeller is driven by the engine and produces a very high rate of fluid circulation. The rate of circulation depends to some extent on the direction and velocity of the fluid which is returned to the impeller by the previous member (usually the reactor).

Hydrokinetic drives are designed with a minimum flow path and the pump, turbine and reactor members are arranged so that the mean flow path of any particle of fluid is approximately circular with a radius of about 2–3 in. In order to obtain the high rates of flow, the bladed members are in the form of annular rings and fluid flows through the blades or vanes set around the whole periphery

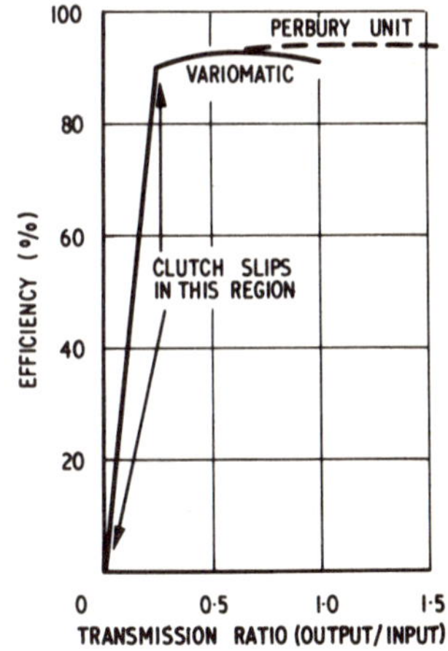

Fig. 6.5 Typical efficiency curves for oil-film drive (Perbury) and variable pulley drive. (DAF)

or circumference of the ring. For the purpose of discussion or analysis it is sufficient to consider that the fluid flows in a circular path whose axis is offset and at right angles to the input or output shafts. Different arrangements of pumps, turbines and other rotary members can be compared with reference to a fluid circuit diagram which can be obtained from a section through the drive and parallel to the flow path.

The turbine and other bladed members are designed so that, when inserted in this rapidly circulating fluid, they deflect the fluid and in so doing a torque is developed on each member. Rotating output members are thus called turbines (see Plate 6.1) and if a member is held stationary it is usually called a reactor member.

The major difficulty with these fluid drives is the fact that with fixed vanes and blades it is possible to obtain maximum efficiency at only one particular set of conditions. It is possible to calculate the various blade angles and shapes so that fluid circulates easily from one member to another with minimum resistance to flow when the turbine is running at, say, 0·8 times the pump speed. At any other set of conditions the entrance and exist angles of the bladed members are not correct and fluid leaving one member will

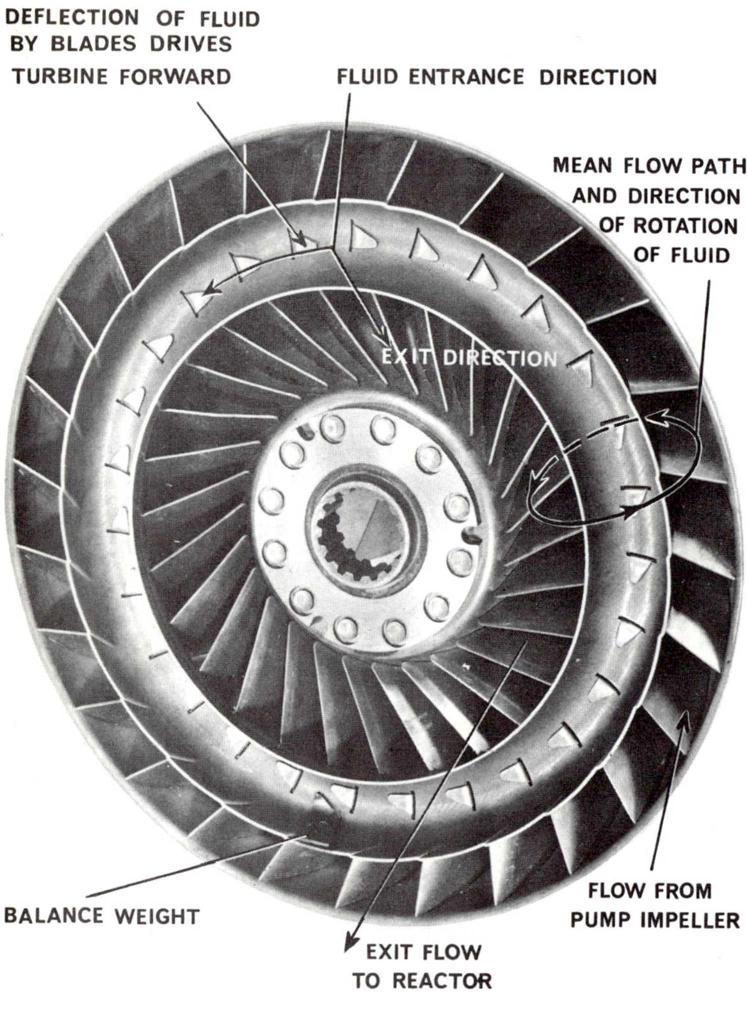

Plate 6.1 Arrangement of blades in a typical hydrokinetic turbine member. Fluid flow is deflected by the blades in this turbine and this deflection develops a torque on the turbine

strike against the side of the blades in the next member instead of entering the member in a direction parallel to the blades. This incorrect entry leads to an entrance loss or 'shock' loss in which some of the forward velocity of the fluid is lost. The fluid also circulates inside and around the flow passage or duct instead of passing through in a parallel and streamline manner and this induces further frictional losses.

The normal operation of the drive at its maximum efficiency still incurs some losses due to the velocity of circulation of fluid and the best efficiency is not likely to exceed 90%. When operating away from the designed conditions, further losses are incurred and the efficiency curve falls off on either side of the maximum efficiency point.

Reaction members seem to present the greatest design difficulties because they experience wider variation in flow velocity and in the direction at which the fluid arrives at the entrance to the reactor. As was seen in Chapter 4 (Fig. 4.1) by mounting the reaction member on a freewheel device (one way clutch) it is possible to obtain an increase in efficiency near the direct drive (1 : 1) ratio. This arrangement is called a converter-coupling and an efficiency of 90% can be obtained for about the top third of the required range of variation. If the maximum efficiency point is moved back to obtain higher torque multiplication then there is a resultant drop in efficiency around the 0·8 speed ratio position since the coupling is only 80% efficient at this point. Consequently the most common hydrokinetic fluid drive is one with best efficiency at the higher speed ratios and with lower torque multiplication. Additional gear ratios are employed to extend the range of torque multiplication.

Many efforts have been made to extend the maximum efficiency region of these fluid drives but the results scarcely justify the complication involved. Several attempts have been made to produce adjustable blades or vanes so as to be able to vary the entrance or exit angles, etc. To adjust the maximum efficiency point (design point) however, it is necessary to alter the angles on two of the rotary members. These adjustable blades not only complicate design and manufacture, but do not achieve all the desired improvement. Pivoting blades can only obtain part of the required design change which requires a three-dimensional change of the ducts and passageways through which the fluid passes.

In any event, it appears to be difficult to achieve a high coupling efficiency as well as a high torque multiplication in the same unit. High efficiency is obtained by employing lower rates of fluid circulation whereas high torque multiplication usually needs much

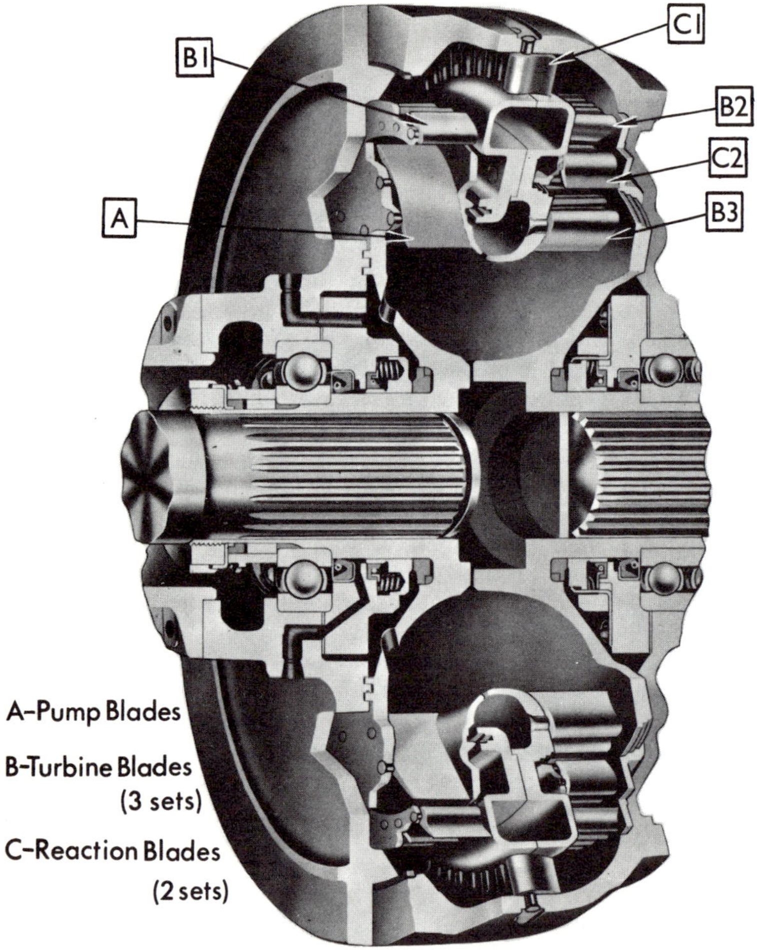

Plate 6.2 Main components of a multi-stage fluid converter

higher rates of flow. For commercial or off-highway applications where high torque multiplications are essential it is necessary not only to achieve high rates of circulation but to increase the number of bladed elements. Thus, in the well-known Lysholm-Smith converter, there are three sets of turbine blades and two sets of reaction member blades. In order to obtain a torque multiplication of 5 to 1, a single turbine is insufficient and it is necessary to pass the fluid more than once through turbine blades. After the fluid has passed through one set of turbine blades it enters a reaction member whose blades collect the fluid and redirect it so that a further turbine member can develop additional torque. In this design there are six sets of blades, one pump member, two sets of reaction blades and three sets of turbine blades (see Plate 6.2).

This type of fluid converter does not operate very well as a coupling because it is inconvenient to release the reaction member and, in general, this converter is combined with a friction clutch which 'locks up' for a direct drive. For passenger cars, and lower torque multiplications, it appears to be more satisfactory to obtain a smaller improvement by dividing the reaction or other members into several elements which can rotate or freewheel as the direction of fluid changes. Thus as seen in Plate 6.3 and Fig. 6.6 it is possible to divide a reaction member into two parts each of which is mounted on a one-way clutch.

As the turbine member accelerates, the resultant direction of fluid arriving at the reaction member may change by nearly 150°. By arranging for this member to be made in two parts, the first part (to the left in Plate 6.3) may freewheel when the fluid direction has changed through say 60°. The fluid drives the reaction member in the forward direction so that there is little change in the direction of the circulating fluid. The rear part of the member is more suited to the new direction of flow and the fluid converter acts as though the front part of the member was removed. The efficiency curve of the converter-coupling now shows three peaks of efficiency (see Fig. 6.6). The successive freewheeling of the two parts of the reaction member prevents the converter efficiency falling too far after having reached the maximum point. This two-part member makes it possible to select two design points before reverting to the usual coupling operation. In this example the coupling performance is assisted by a lock-up clutch which enables the coupling operation to reach direct drive (1 :1 ratio) and the 100% efficiency point. In practice, of course, there are auxiliary power supplies which absorb some of the engine's power so that the transmission can never attain 100% overall efficiency.

This two-phase converter-coupling has an improved efficiency

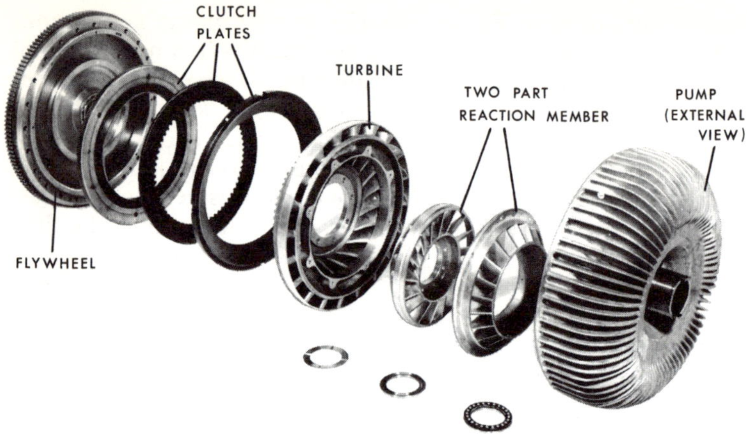

Plate 6.3 Main components of a 2-stage fluid converter

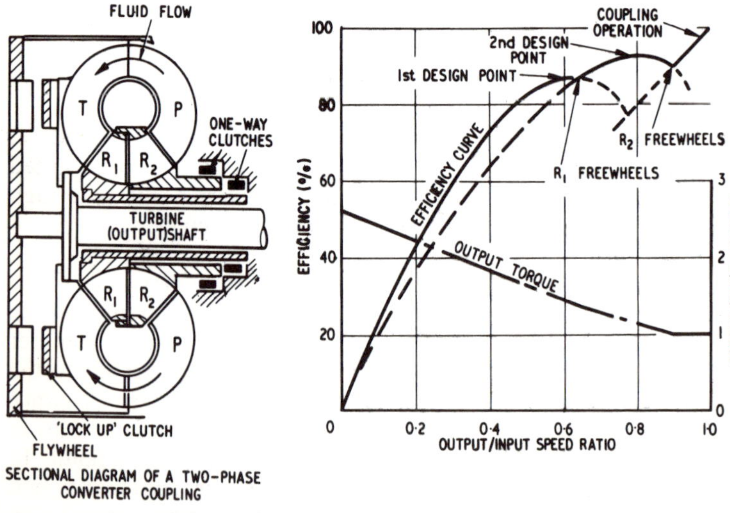

Fig. 6.6 Sectional diagram and performance characteristics of a 2-phase converter-coupling

and a higher torque multiplication than a single-phase version (see Fig. 4.1) but the output torque is still not sufficient for all road operation and extra gear ratios are needed in the form of an automatic gearbox.

Since it appears that hydrokinetic fluid drives will always require some assistance from gearing, one of the more recent innovations has been to connect these freewheeling members to gearwheels in the associated gearbox. Thus instead of freewheeling idly, these additional bladed wheels are able to deliver some additional power. This technique is a feature of some Swedish and American transmissions and the principle is illustrated in Fig. 6.7 which shows the

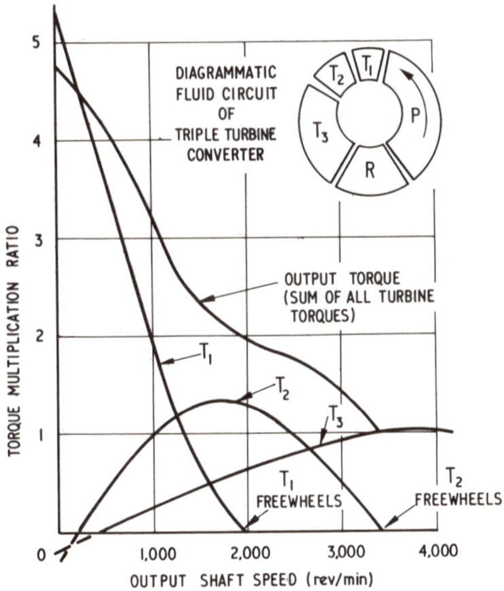

Fig. 6.7 Performance characteristics of a triple turbine transmission.

overall performance of the 'Chevrolet Turboglide' and Buick Triple-Turbine transmissions (see also Plate 6.4).

As seen in the diagram of the fluid circuit there are five bladed wheels in the flow path. In this example the turbine is divided into three parts each connected by concentric shafts and able to rotate without reference to other members. Each of the three turbine members is connected by gearing to the output shaft and the gearing is combined with one-way clutches so that the sum of all the turbine torques is transferred to output. It will be noted from Fig. 6.7 that each turbine develops a maximum torque at some selected output speed and that above or below this speed the torque

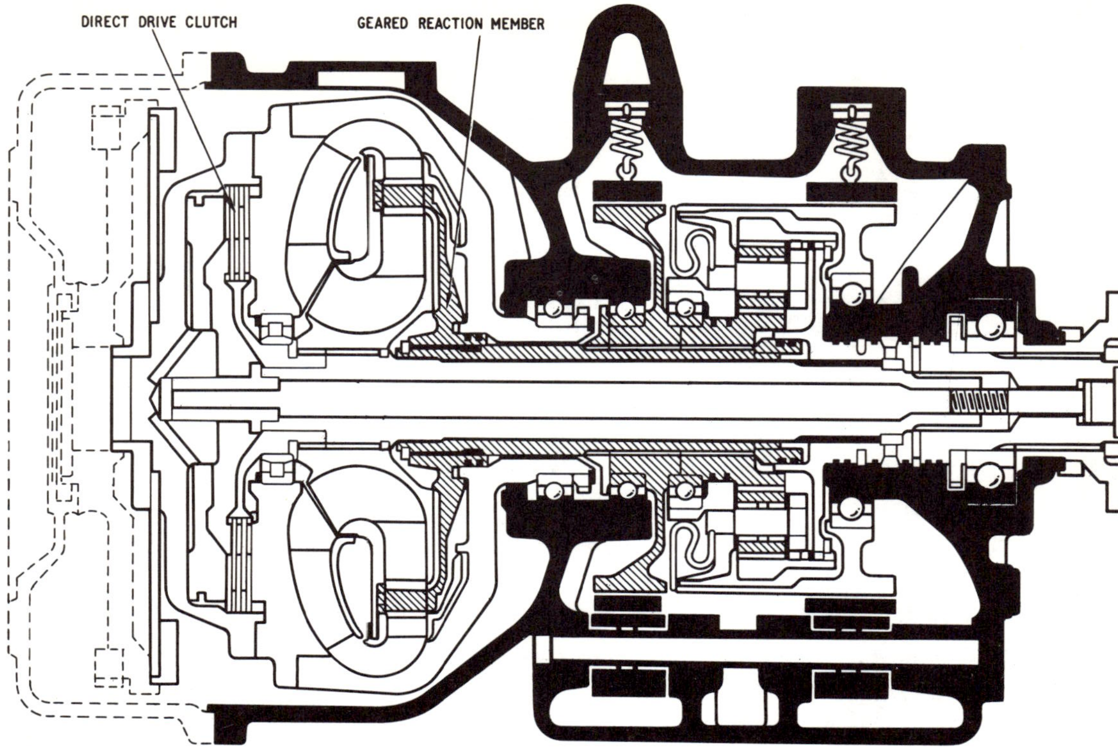

Fig. 6.8 *In the Bristol-Siddeley SRM converter and the triple turbine transmissions, the bladed converter members are connected to gears in the epicyclic sets so as to add, and multiply, the torques developed on the turbines and reacters*

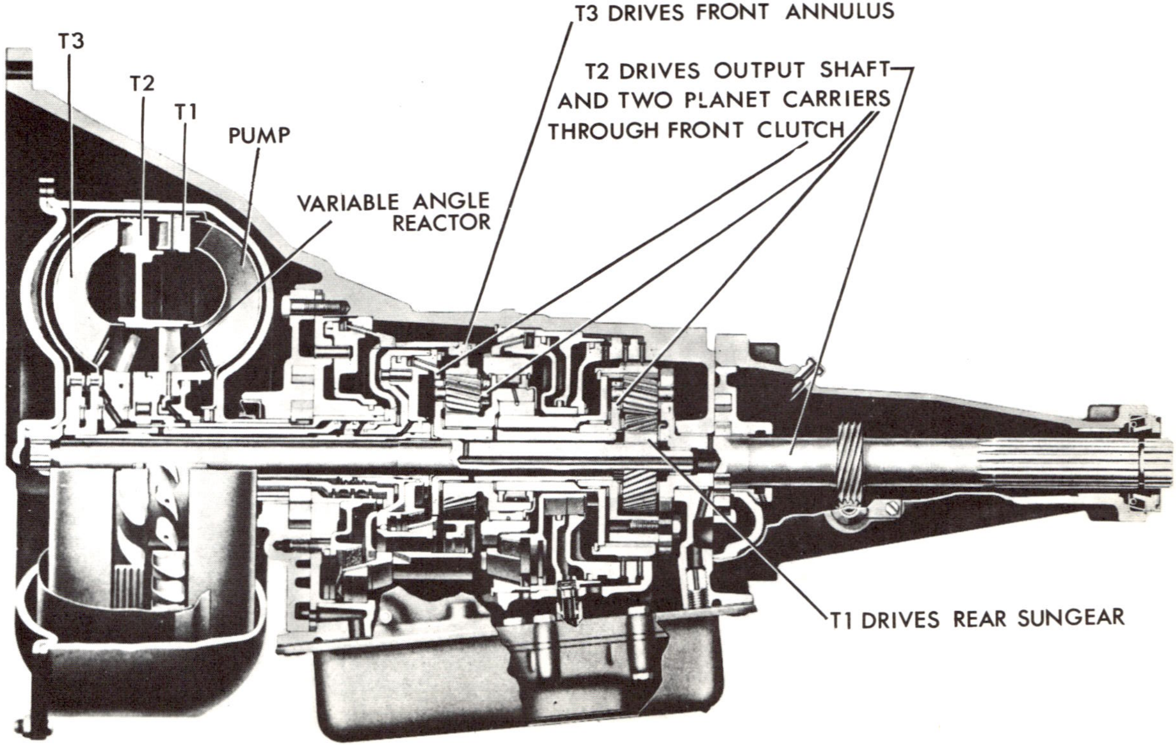

Plate 6.4 The Chevrolet Turboglide. (By courtesy of General Motors Ltd.)

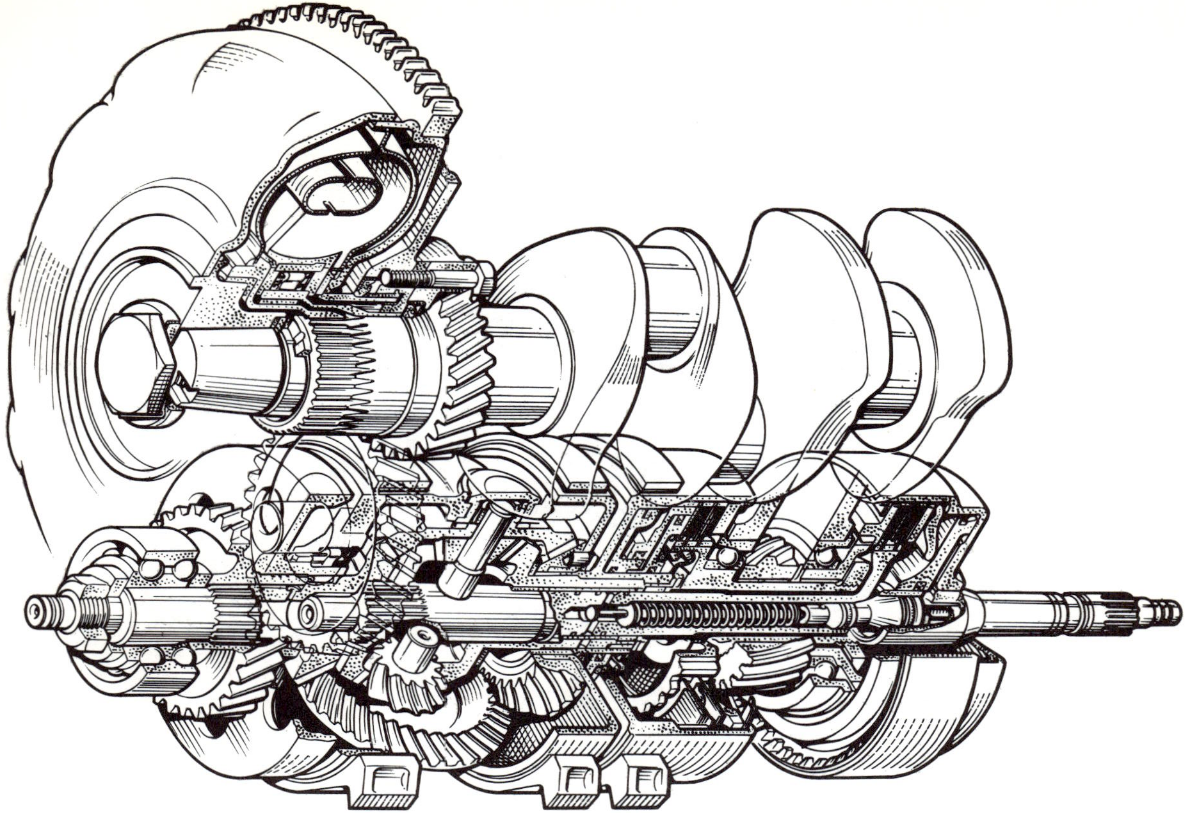

Fig. 6.9 The BMC-AP fully-automatic transmission for Mini and 1100 cars

falls off. Turbine T_1 is designed to produce a high starting torque and its output is multiplied 2·86 times by gearing. Turbine T_2 is designed for an intermediate speed range, and gearing increases its torque by 1·54, whilst turbine T_3 is employed for the normal coupling operation and is directly connected to output. Turbines T_1 and T_2 are released by the one-way clutches when their torque falls to zero. When starting, these two supplementary turbines develop a small negative torque, i.e. act as pumps to give additional circulating velocity to the fluid. Fig. 6.8 shows a similar principle.

In this type of hydrokinetic fluid transmission, all possible use has been made of the torque developed in a series of bladed wheels operating over a speed range in a fluid flowing at high velocity. Even so, the total output torque is still not much improved on a simple converter coupling combined with three auxiliary gear ratios. The output torque from the triple turbine shows the three separate efficiency peaks as each turbine reaches its maximum conditions. Although a smooth transition through its operating range is achieved, the overall complication is probably greater because of the more complex converter design and increased number of blades, wheels and bearings, etc.

For British and European use, where initial and operating costs must be low, it does not seem that the hydrokinetic fluid drives will penetrate very far. Although as in the Mini-Automatic (Fig. 6.9) they operate smoothly, and have good starting characteristics, these drives apparently require at least as many auxiliary gear ratios as the gearbox they replace. Consequently, it is difficult to justify the additional cost of this fluid drive. Smaller vehicles, with more need of torque multiplications, will be finally satisfied only by a transmission having efficient and wide range of ratios, and hydrokinetic fluid drives appear to be too expensive a solution for full scale production applications.

The more recent AP Mk III transmission is intended for more powerful vehicles, but by rearrangement of the gearing and clutches, has been considerably lightened in weight. The new version is 4 lb lighter than the earlier small car versions and comparison of the two alternatives emphasises the problems in achieving an economic solution for the smaller car. (see Fig. 6.10)

6.3 HYDROSTATIC FLUID DRIVES

In hydrostatic fluid drives, power is transmitted by the movement of fluid under pressure. In such transmissions it is possible to employ very high fluid pressure so that the flow of fluid is very low and almost negligible by comparison with the hydrokinetic type.

A hydrostatic fluid drive consists of a pump, which converts the torque and rotation of a mechanical shaft into a flow of pressurised fluid, combined with a hydraulic motor which converts the fluid flow back into a rotating torque on an output shaft. The pump and motor units are usually identical in principle but may vary in size and fluid displacement. Many pumps and motors are of the piston or plunger type in which a series of pistons are reciprocated in a closely fitting bore by rotation of the drive shaft. These pistons are fitted into a cylinder block which contains ducts to collect fluid from each piston and deliver it to inlet and outlet ports. The pistons may be arranged radially or axially concentric with the drive shaft and are known as the radial-piston or axial-piston versions respectively. Vane type hydraulic units are also employed for power transmission but usually operate at lower pressures.

Hydraulic units employed in hydrostatic drives must be of an adjustable type so that the volume of fluid displaced during each revolution of the drive shaft can be varied (see Plate 6.5). Such units are known as the variable delivery type and the vane and piston type units lend themselves more easily to the variable delivery duty. Radial-piston units and vane type hydraulic equipments have their adherents but axial-piston units have been subjected to more development and variation.

Hydraulic power transmission by axial-piston units has been available for at least 50 years and the earliest applications were in shipping. Improvements achieved by various aircraft development work has resulted in a considerable reduction in size of equipment for a given power. Much higher operating speeds and pressures are now possible and hydraulic equipment is growing rapidly in application to many industrial purposes. Not unnaturally, many efforts are being made to employ hydraulic equipment for vehicle transmission although this application has extreme requirements for performance and in the cost of the device.

Transmission of power by means of fluid at high pressure offers a new range of possibilities in vehicle layout because hydraulic pipes can replace mechanical shafting. Hydraulic motor(s) may be placed close to driving wheels or axles and fluid may be supplied by piping from a hydraulic pump which may be situated with the vehicle's engine at any suitable position. Although individual hydraulic motors fitted in each wheel of a vehicle may be advantageous for specialised or off-highway vehicles, it soon can be shown that a single hydraulic motor driving through a normal axle is a more compact and cheaper transmission. If hydraulic equipment becomes less expensive than gearing, then the wheel motor concept becomes more attractive.

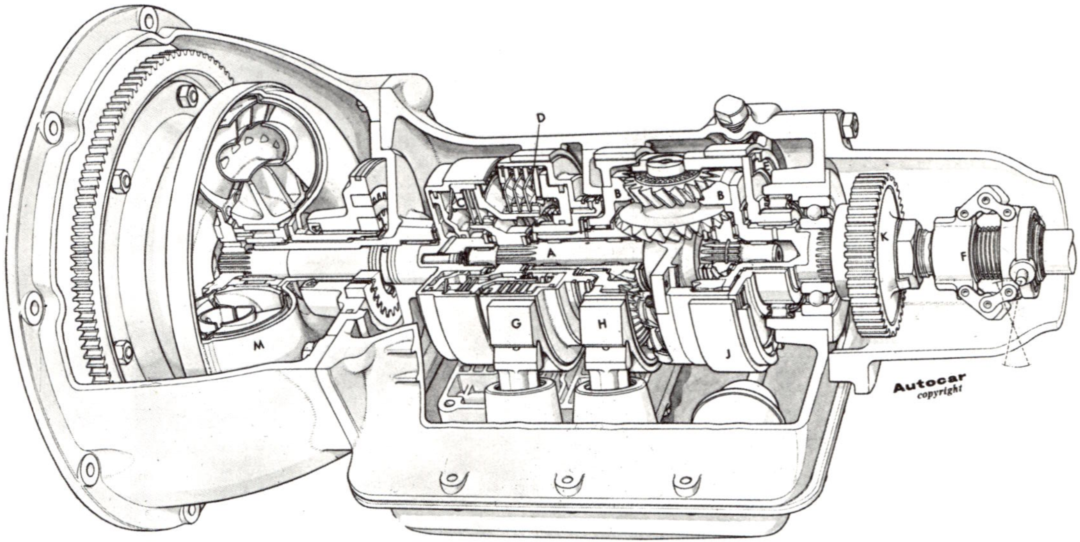

Fig. 6.10 *A cutaway drawing to show a typical arrangement of the internals of the AP Lark III. Note that the governor is now fitted to the output shaft. A, Input shaft; B, Mair gear cluster; C, Forward clutch; D, Top and reverse clutch; E, Sprag clutch; F, Governor; G, Thrust gear brake band; H, Second gear brake band; J, Reverse gear brake band; K, Parking lock; L, Main oil pump; M, Torque convertor*

For most of the applications mentioned at the beginning of this chapter, a hydrostatic fluid drive will consist of a single pump and motor, situated close together in a casing which takes the place of a normal gearbox. As seen in Fig. 6.11, another type of hydrostatic drive consists of two variable delivery units. An input shaft drives a

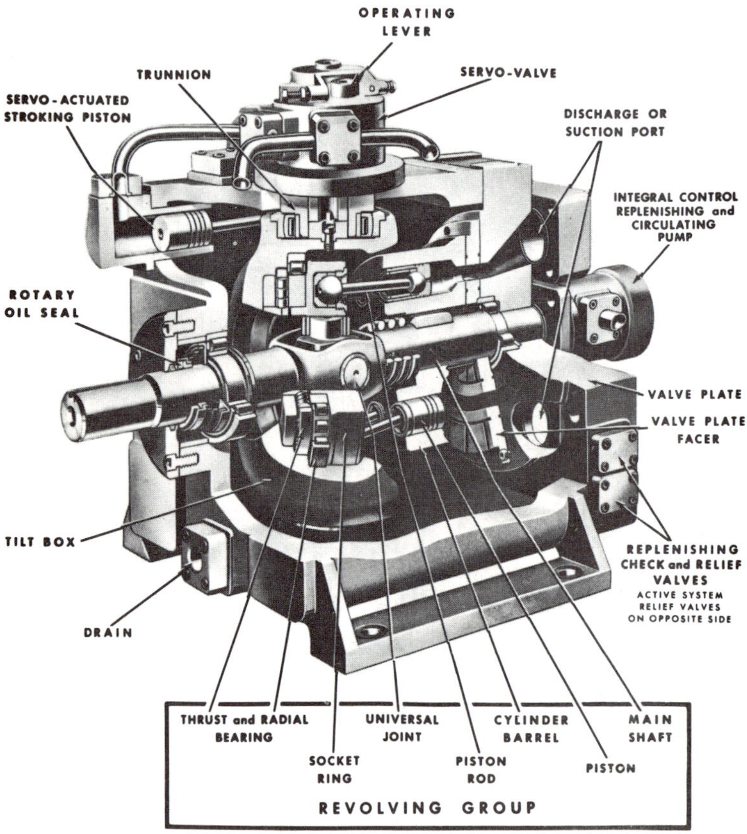

Plate 6.5 A sectional view of a typical hydrostatic pump or motor unit. (By courtesy of Dowty Group Ltd.)

socket ring and hence, a cylinder block containing a number of pistons and this cylinder block can be adjusted to an angle within a range of about $\pm 20^\circ$ of a neutral position. When the block is tilted, rotation of the input shaft causes the axially disposed pistons to move up and down their bores in the pump cylinder block. Connecting-rods on the end of each piston are held in position by a socket ring, and rotation of the tilted block moves the pistons in their bores. Each piston completes one up and down stroke during one revolution of the input shaft. The volume of fluid displaced by each piston is dependent on the tilt angle of the cylinder block and this angle is increased or decreased to bring about a corresponding change in the volume of fluid discharged from the pump.

The fluid discharged from each piston is collected from the end face of the cylinder block by an outlet port in a stationary port plate which seals this face of the block. This port plate also contains

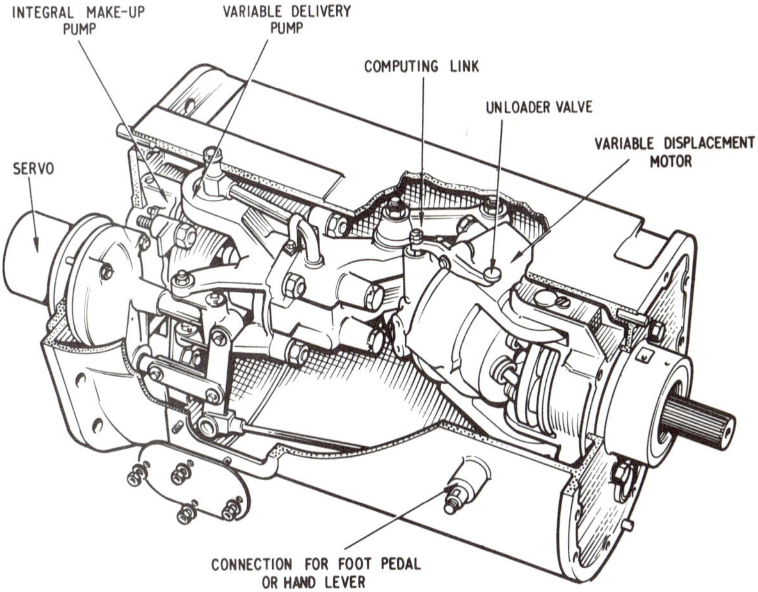

Fig. 6.11 An arrangement of two variable hydrostatic units to form a variable transmission

an inlet port which supplies fresh fluid to each piston bore when each piston is moving on its downstroke. The pump port plate connects with a motor port plate by means of pipes in the transmission casing and fluid flows from pump to motor at high pressure through the pipes and connecting ducts. Return fluid from the hydraulic motor returns at low pressure to the inlet port of the pump when it starts another cycle of operation.

A hydraulic motor of the axial piston type is usually similar to its pump except that for most applications where torque multiplication is needed, the motor is usually larger in size and fluid displacement than the pump. As seen in Plate 2.5, the motor has similar cylinder block, pistons, slipper pads, camplate and other items as in the pump. The application of fluid under pressure to the inlet port of this motor causes the cylinder block to rotate driving the output shaft which can thus do useful work.

By employing variable delivery hydraulic units it is possible to obtain a wide range of output ratios. Thus, if the pump unit is adjusted so that it accepts and discharges 5 in^3 of fluid at pressure for each revolution of the input shaft, then it will be possible to choose a hydraulic motor whose volume can be varied from 20 in^3 per revolution down to a minimum quantity. In such a case the pumpshaft would rotate 4 times to each revolution of the motor shaft (4 : 1 reduction ratio) and by reducing the angle of the motor camplate, the motor displacement can be reduced progressively to any value until at a quarter of the maximum camplate angle, it will require only 5 in^3 of fluid for each revolution of the output shaft (1 : 1 ratio—direct drive). In this way, a fully variable ratio transmission with a 4 : 1 range is obtained.

Further reduction of the motor camplate angle would send the transmission into overdrive range, but the same effect could also be obtained by increasing the displacement, i.e. camplate angle of the pump unit. The use of two variable units enables a very wide range of output ratios to be obtained, and since the camplate can be moved to either side of its zero position, similar ratios in a reverse direction are easily obtained. In practice, small camplate angles—below 5°—are not often used, but even so, a 3 or 4 : 1 variation with each unit seems practicable, and a transmission having a 10 : 1 range of variation should thus be possible.

In Fig. 6.11 the axis of the cylinder block is tilted relative to the socket ring, and the angle of tilt has the same effect as camplate angle. The camplates or cylinder blocks are adjusted by small servo-pistons and linkages to positive negative angles of tilt about the vertical, to obtain the required transmission ratio. These servo-pistons are moved by a control cam which moves two control-valves—one for each piston—to admit a low pressure supply of hydraulic fluid to one side or other of their pistons. This low pressure supply is also used as 'boost' which ensures that the low pressure ports are maintained at a low pressure (about 100 lb/in^2) to make up for any leakage losses and to prevent air leaking into the fluid circuit. This 'boost' pressure must be supplied from a separate small pump unit and is also used to circulate fluid for lubrication or cooling.

A hydrostatic transmission has advantages over a mechanical equivalent because the hydraulic version can be easily and progressively connected or disconnected by means of a by-pass valve which permits fluid to flow from a high pressure port to a low pressure port. The hydraulic transmission can be disconnected partly or completely in this way and does not require any starting device. A two-way relief valve in a hydraulic transmission can be used to prevent overloads which may occur for any reason during normal driving or over-run conditions, and this is a precaution not easily achieved in a mechanical transmission.

There are a large variety of arrangements, patents, and other differences in the design of axial piston machines although the basic pumping principle is identical—see Fig. 6.12—in all machines. Although there are many possible variations in the manner in which the piston connects with the camplate, and in whether the cylinder block or the camplate is held stationary, the useful force T is always proportional to the camplate angle, and the applied pressure load P as follows:

$$T = \tfrac{1}{2}P \sin 2\theta \tag{6.1}$$

From this relationship it would appear more advantageous to employ much larger camplate angles than the relatively small ones previously mentioned. However, as θ increases, the useful force T appears as a side load or torque on the working plunger or connecting rod. Problems of wear are acute at these points in the hydraulic unit so that the state of development of these axial-piston designs may often be judged by the maximum camplate angle which is permitted. Whilst it seems comparatively easy to operate up to camplate angles of 15°, there are only a few versions which can be permitted to operate at angles of 30° at high speed and high pressure.

Although earlier designs attained high camplate angles with the connecting rod versions, there has been a noticeable tendency to adopt the more compact slipper pad technique. The slipper pad is usually pressurised to minimise the mechanical load between the pad and the camplate, but owing to variation of this load due to changes in angle, it is not possible to balance the load completely. There is a friction force developed between the piston end or pad due to its movement around the camplate and this force alters the direction of the camplate load R. This friction force F acts on the piston end in the same direction as the camplate motion and gives a relative increase in camplate angle when pumping and a corresponding decrease in this angle for the motoring function.

In some pump designs the pistons are inclined at an angle to the

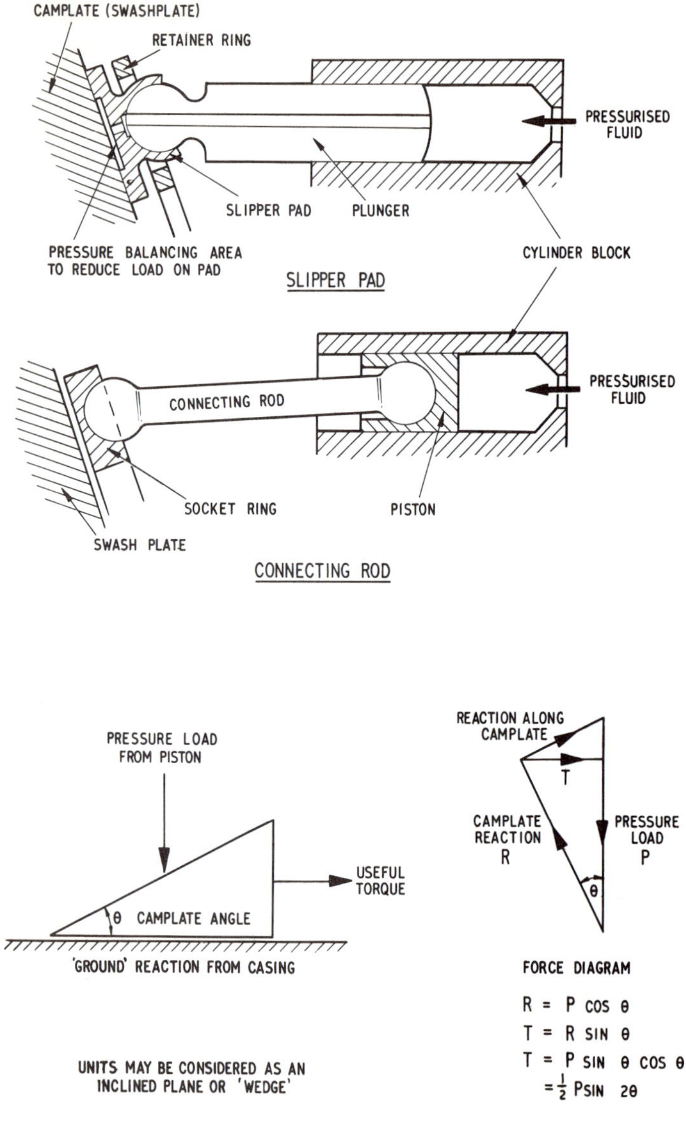

Fig. 6.12 Basic pumping mechanism for hydrostatic piston machines and equivalent force diagrams

truly axial—i.e. parallel to shaft—position so as to distribute the side loads more evenly over a working range. The curves given in Fig. 6.13 show the areas of performance, i.e. speed and torque, which are executed at a particular efficiency as stated in the enveloping curves. These characteristic curves are typical of most pressure-hydraulic machines whose efficiency is the product of a volumetric efficiency and a mechanical efficiency. Volumetric efficiency increases at each value of torque with an increase in speed, because the loss of fluid which leaks from port plates and through small working clearances depends mainly on the applied torque and the

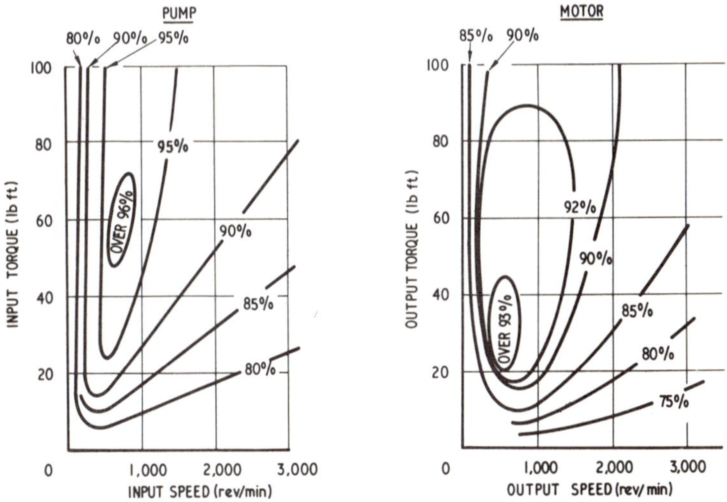

Fig. 6.13 Overall efficiencies of an axial piston unit for pump and motor duties

lost quantity becomes a decreasing proportion of the volume pumped at higher speed. Mechanical losses are mainly due to friction and mechanical efficiency falls as the speed of the machine increases. The overall efficiency is a combination of these two effects and peak efficiencies are obtained at comparatively low speed and torque.

When a pump and motor are combined to form a variable ratio transmission its overall efficiency is again the product of the pump and motor efficiencies. The best efficiency that could be expected from the examples previously shown is $93\% \times 96\% = 89\%$. However, when the two charts are superimposed, the two areas do not overlap and the pump efficiency of 96% will fall in the 92%

area for the motor. Similarly, the 93% efficiency of the motor corresponds to 95% (or less), areas of the pump performance, and the best possible performance is seen to be a small range at 88% efficiency.

Consequently, when combining two hydrostatic units to form a variable ratio transmission it is important to combine the best areas of performance in each unit. Most designs attempt to keep the motor unit at maximum tilt angle since the motor is working in a lower efficiency region than the pump, but it eventually becomes necessary to reduce the tilt angle in the motor if a sufficiently wide range of variation is required. In the example shown in Plate 6.6 the tilt angle of two variable delivery units was varied by means of a computing link which connected the moveable cylinder blocks in a pre-determined sequence. By positioning the pump and motor at a given distance and employing a mechanical connecting link of the correct length, servo adjustment of the pump tilt angle caused the motor tilt angle to move as shown in Table 6.2 below:

Table 6.2

Torque Multiplication Ratio	*Pump Tilt Angle* (Degrees)	*Motor Tilt Angle* (Degrees)
6	2	25
5	11	20
4	13	19
3	15	$17\frac{1}{2}$
2	22	12
1	24	9
(Reverse) −5	− 12	23

Some of these positions are illustrated in Plate 6.6

Despite the ingenuity of this and other systems in combining the best areas of performance of each unit, there is still little prospect of obtaining much more than the 88% previously calculated. In fact, when the transmission is mounted in its case with extra seals, bearings and boost pump losses to be considered, the final performance is not as good as this figure. Fig. 6.14 shows an efficiency band which is typical of the best performance of hydrostatic drives. Poor matching of pump motor efficiencies could lead to much lower values.

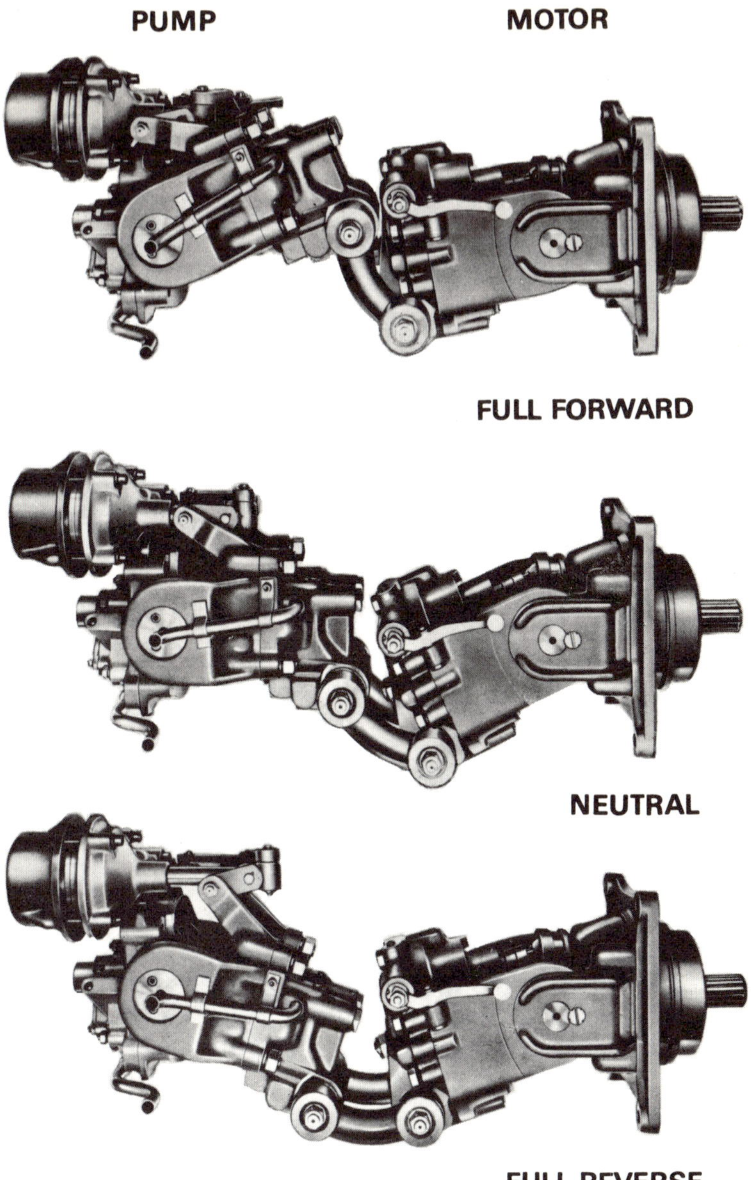

Plate 6.6 A 'Computing' link is employed to combine the best efficiency regions in pump and motor units for a variable ratio transmission. (By courtesy of Dowty Group Ltd.)

Clearly, such performance and efficiencies are not yet adequate for road vehicle use. Although the torque multiplication of these drives is quite adequate for passenger cars, there are limitations in maximum speed of operation and in efficiency which have not yet been overcome. Development work continues and in time

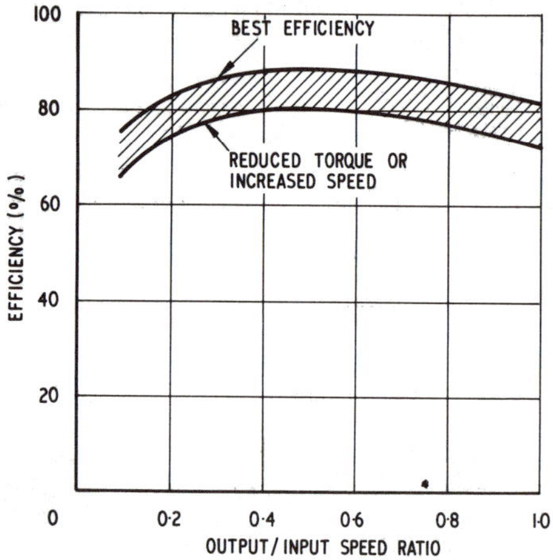

Fig. 6.14 Probable working range of efficiency for a hydrostatic drive

further improvements may bring this hydrostatic drive into wider use. At the present time, this type of fluid drive is extremely popular for various industrial drives, machinery, etc. and is beginning to find application in off-highway and specialised vehicle duties.

6.3.1 VANE TYPE HYDRAULIC EQUIPMENT

Many applications in which axial-piston equipment is being employed can be satisfied by hydraulic equipment operating at lower pressures, and various types of vane pump are less expensive and less bulky for such purposes. The simple sliding vane type of equipment has few parts and can be made in variable delivery versions, for pump or motor duties. Several different designs are in use for the same applications as those employing axial-piston equipment including off-highway tractors and crawlers.

Vane pumps operate at lower pressures than axial-piston equivalents although pressures of 2,000 lb/in^2 have been employed. The advantage of a vane type equipment is the large volume of fluid delivered by it. For a given size or weight of pump, the vane type more than makes up for any pressure limitations by delivering a much larger volume of fluid. Compared on a power/weight basis, vane type equipment is better than many other forms of hydraulic equipment.

The development of vane-pumps and motors has not advanced as far as the axial-piston equipment but Fig. 6.15 shows an illustration of a typical sliding vane pump together with a slipper-vane variation designed by the author. In either version, rotation of an input shaft causes the vanes or slippers to rotate inside a guide ring which is mounted eccentric to the input shaft. The vanes slide in or out relative to the rotor as they move around inside the guide ring and the spaces between the vanes (slippers) increase and decrease during each revolution. Inlet and outlet ports are positioned so that fluid enters these spaces whilst the space is increasing in volume and is discharged by the contraction of these spaces.

Adjustment of fluid delivery can be made by sliding the guide ring towards the concentric position. When concentric, the vane spaces maintain a constant volume and no fluid enters or is discharged from the pump. If the guide ring is moved over-centre, then the identity of the inlet and outlet ports is reversed and fluid is pumped in the reverse direction. Inlet and outlet ports are symmetrically placed about a bisecting axis and movement of the guide ring must always take place along this line.

The efficiency of vane pumps and motors is improving and approaching that of axial-piston equipment. Volumetric efficiency of vane type equipment is low owing to the leakage across the thin vane, but this is improving in the case of the thicker slipper-vane version. Mechanical efficiency of this type is high and the overall efficiency is maintained over a wider range of speeds than piston and plunger types. By restricting the operating pressures to 750–1,000 p.s.i. so as to maintain a good volumetric efficiency, then the vane type of equipment can reach high overall efficiencies. The friction forces, due to rotation of the vanes inside the ring appear to work in the opposite sense to the axial-piston equivalent so that vane motors, especially the slipper type, appear to have a higher efficiency than the pump version.

6.4 HYDROMECHANICAL SHUNT TRANSMISSIONS

Although hydraulic machines have a much higher mechanical efficiency than many other machines, they cannot match the very high efficiencies which can be attained by mechanical gearing. One

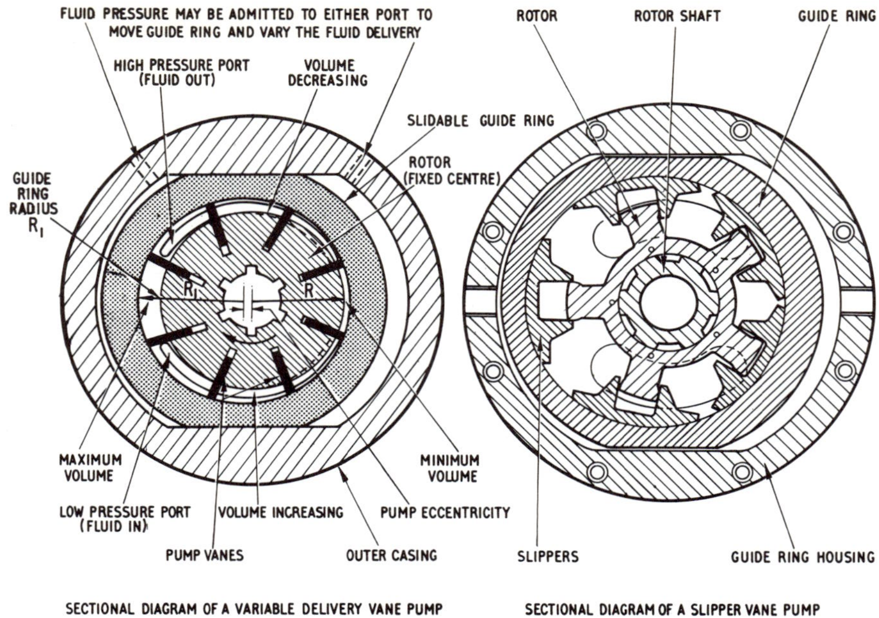

Fig. 6.15 Variable delivery vane pumps

of the ways in which the efficiency of hydraulic power transmission equipment is being improved is by arranging for part of the power to be transmitted by mechanical means—either by gearing or by structural members, casings, etc. The term hydromechanical refers to a combination of hydraulic and mechanical transmissions and the term 'shunt' as in railway or electrical terminology, indicates alternative paths.

Thus, a hydromechanical shunt transmission is one which provides alternative paths for power flow from input shaft to output, one path transmitting power through hydraulic medium, the other by mechanical means. In these arrangements it is possible to combine the variable ratio properties of a hydraulic drive with the higher efficiency of a mechanical drive in order to obtain a variable ratio drive of improved efficiency. In a hydromechanical shunt transmission it is necessary to divide the incoming power, so as to send a proportion through the hydraulic and mechanical transmissions and then to re-combine the two outputs in a single output shaft.

6.4.1 CASING REACTION SYSTEM

Although the process of dividing and recombining the power flows may sound a complicated procedure, it can be done in a very simple way with a hydrostatic drive. In the casing reaction system, no additional parts are necessary and the components of the drive are merely rearranged in a different manner. Mechanical drive in the casing reaction system is achieved—as its name implies—by transmitting a mechanical torque through the pump and motor casings.

The division of power flow is obtained in the pump unit whose drive shaft is connected to the input shaft in a normal manner. As soon as this drive shaft is turned, a reaction torque is developed on the pump casing which normally is held stationary or bolted to the vehicle or an engine casing. If the outlet from this pump was blocked in some way, the input shaft would not rotate although very high internal pressures might develop and, of course, an equal but opposite reaction torque on the casing would be present. If the pump casing was free to rotate in bearings then when the outlet was blocked, the whole assembly, input shaft, rotor and casing would rotate together and there would be no movement of fluid.

In the casing reaction arrangement, the pump and motor casings are free to rotate and mounted in bearings in another transmission casing. When the pump outlet is blocked, rotation of the input shaft will cause the pump and motor casings to rotate with it. In this condition, the whole pump motor assembly is in a 'locked-up' condition and must rotate with the input shaft. It is possible, however, to obtain an output drive from these rotating casings which—with a blocked pump outlet—are capable of transmitting an output torque and this is what is done in this arrangement. The pump and motor casings are connected directly to the output shaft and when the pump is blocked, the whole assembly rotates as a direct mechanical drive for a 'top gear' condition.

In this shunt arrangement, the pump and motor are able to act as a transmission in a normal way but the whole assembly will be rotating at the same time. The pump outlet is blocked by adjusting the hydraulic motor to zero delivery so that it cannot accept or deliver any fluid—regardless of whether the motor shaft is rotating or not. In fact, the motor shaft is fixed to the stationary outer transmission casing, so that in the locked-up (top gear) condition the motor casing rotates around its stationary motor shaft. Since the motor delivery has been adjusted to zero, the casing rotates quite freely about this shaft.

To increase output torque in this shunt transmission, the fluid delivery of the motor is adjusted to a given setting and a flow of fluid will be needed for the motor casing to rotate around its shaft. This fluid comes from the pump which is now able to deliver fluid and the pump and motor casings rotate at a speed lower than input so that the pump delivers exactly the right amount of fluid required for the casings to rotate around the stationary shaft.

Thus, as seen in Fig. 6.16 which shows a practical example of such a transmission, the output shaft receives the casing reaction torque jointly from the pump camplate and the motor cylinder block. Now the casing reaction on the pump unit is equal to, and acts in the same direction as, the input torque. The hydraulic motor casing will develop a torque equal to the hydraulic drive ratio. As a result, if the hydraulic motor is adjusted to deliver three times the quantity of fluid delivered from the pump then the motor casing torque will be three times the input torque (pump torque). Hence the output shaft shows a total torque of four times input, i.e. a gain of one additional input torque obtained from the casing reaction torque of the pump.

By recombining the conventional pump and motor of a hydrostatic drive in this way it is possible to increase the torque multiplication ratio by one, and to obtain the benefit of increased efficiency which comes from the mechanically transmitted pump reaction torque. Thus if the motor unit is adjusted to zero delivery, the pump is unable to deliver fluid and carries the pump casing

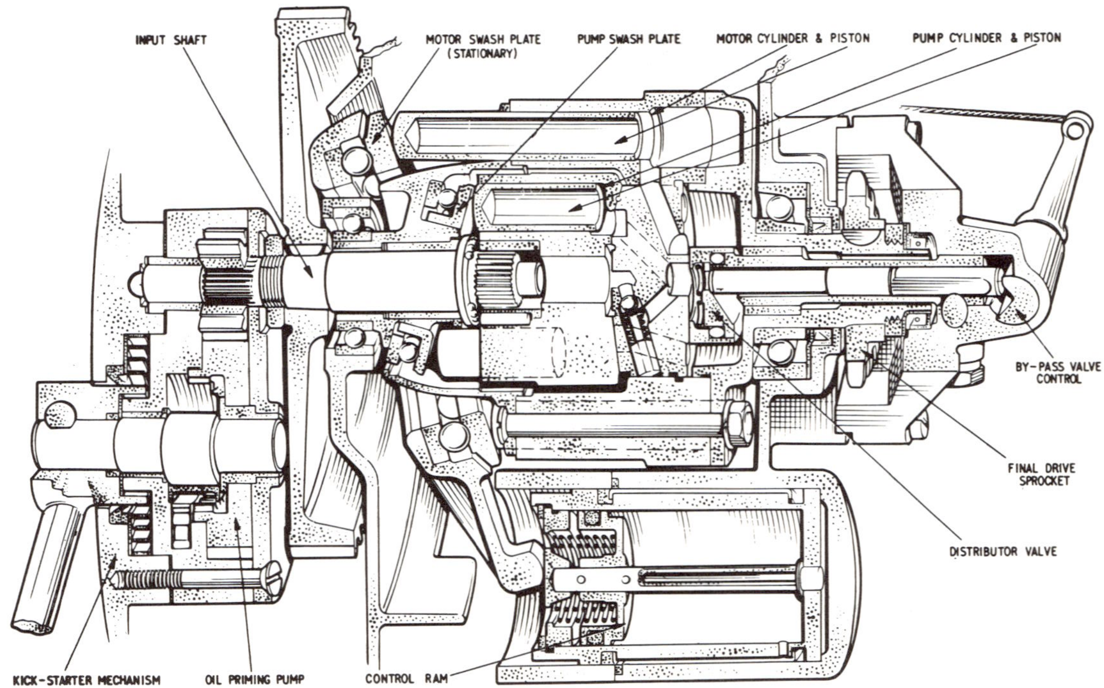

Fig. 6.16 The Badalini transmission for motor cycles and scooters is a shunt arrangement of an axial-piston transmission. The input drive is connected to the pump shaft, but the output sprocket receives the combined torque from the pump camplate and motor cylinder block which are both connected to the output shaft. Reaction torque is obtained from the stationary motor swash plate. Larger versions of this transmission have been successfully run in cars and larger vehicles

with it around the input shaft. Since the pump casing is, effectively, the output shaft, the complete transmission is locked up into a mechanical drive direct (apart from small leakage loss and the motor shaft friction). As the delivery of fluid from the motor unit is increased the output shaft speed decreases but its torque is increased by the amount of torque developed in the hydraulic motor. By increasing the delivery of fluid from the motor, the transmission develops further torque until the maximum delivery setting is reached.

This shunt arrangement has only slightly different requirements for control as compared with a 'straight hydrostatic' version. The motor unit is varied from zero delivery to maximum delivery to obtain the full range of torque multiplication, whereas the hydrostatic version will not use the zero delivery setting and motor delivery will never be much less than half-maximum delivery. In

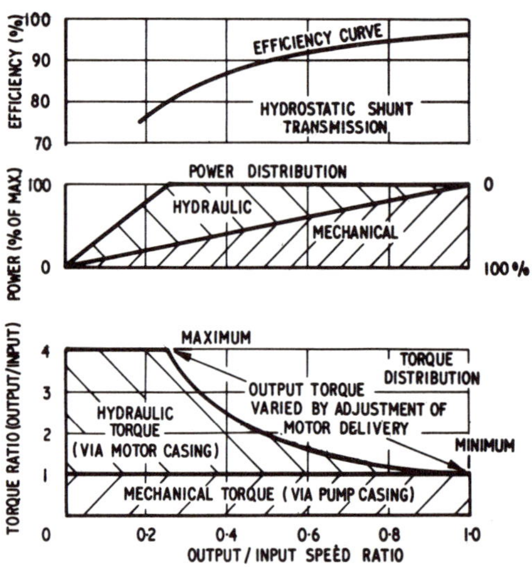

Fig. 6.17 Distribution of torque and power, and a typical efficiency curve for a 'Casing Reaction' hydro-mechanical shunt transmission

both versions, however, it is usual to reduce the delivery of the pump unit in order to extend the range of torque multiplication and the pump or motor camplate will be tilted overcentre for reverse operation and for overdrive ratios. The benefit of the shunt arrangement is illustrated in Fig. 6.17 which shows how the mechanical torque and power are distributed and give rise to an improved efficiency for at least the higher speed ratios. In these diagrams a pump is combined with a motor having three times the fluid delivery of that pump to obtain a casing reaction/shunt transmission giving a 4 to 1 torque multiplication. It can be shown that if the motor delivery is set to 'n' times that of the pump then the output speed ratio 'x' is given by:

$$\text{Output speed, } x = \frac{1}{1+n} \text{ . input speed} \qquad (6.2)$$

It will be noted from these diagrams that a mechanical torque equal to input is always transmitted to output. When the vehicle is started the hydraulic drive transmitted the majority of the engine's power, but as the speed ratio increases, this mechanical torque becomes more important until at a speed ratio of one (direct drive) the drive is all-mechanical. Consequently the efficiency is high—about 95% for top gear—but falls at lower speed ratios to values more typical of a hydrostatic drive.

6.4.2 DIFFERENTIAL GEAR SHUNT SYSTEMS

It is possible to obtain a shunt transmission by combining the hydraulic drive with a differential gear—usually of the epicyclic type, see Fig. 6.18. Mechanical power is transmitted at high efficiency by the gearing but a proportion is transmitted through the hydraulic drive and returns, after conversion to the output shaft. By adjusting the ratio of the hydraulic drive, the output shaft speed may be varied over a small range. If the variable range is extended over too wide a range the proportion of power transmitted by the hydraulic drive becomes greater than the input so that this arrangement is limited to applications where a small range of variation is required. Thus, although this system has been employed for industrial drives, pumps, fans, etc., of very high power it is of less use for automobile transmissions.

The graph shows how a simple epicyclic gear would be combined with a hydraulic drive. Mechanical power is transmitted through the gearing to the output shaft on which the hydraulic motor is mounted. The sun gear in the epicyclic set drives the pump shaft which is mounted concentrically outside the output shaft. The basic principle is shown as the 'box' diagram in which the differential gear divides the input power into two paths, one mechanical, the other hydraulic. Power is converted in the hydraulic drive and is recombined at the point on the output shaft where the

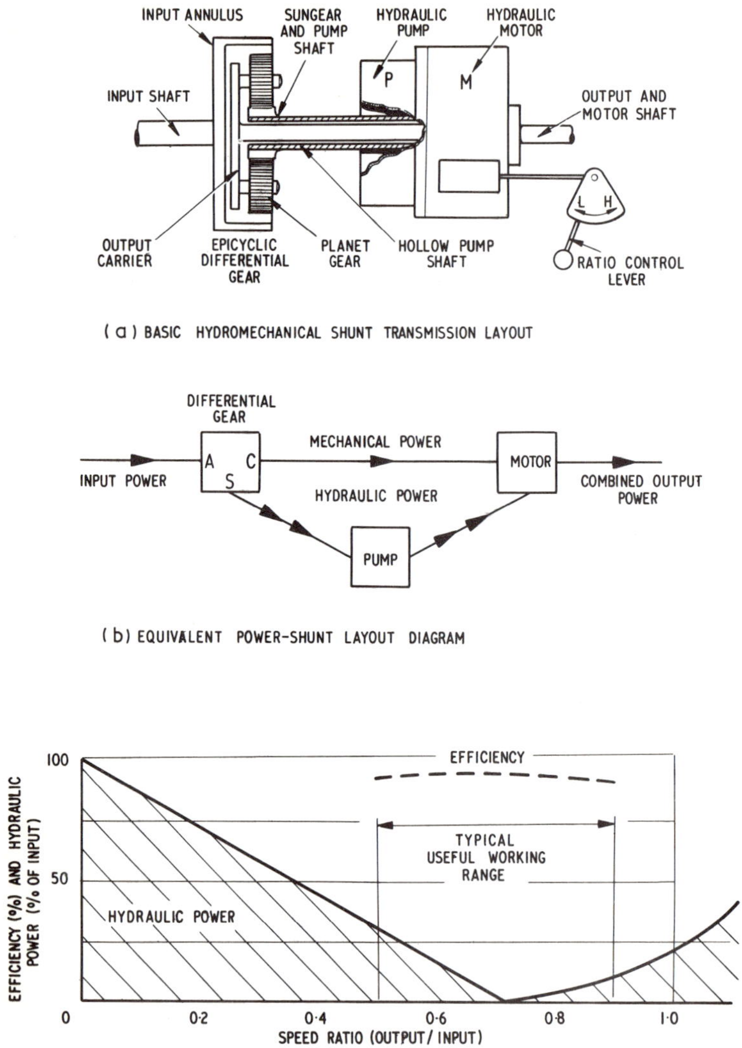

Fig. 6.18 Hydromechanical shunt transmission employing differential gearing

hydraulic motor is mounted. As seen in the graph, the drive is all-mechanical at one point and a high overall efficiency is obtained around this point. At lower transmission ratios the hydraulic power increases and the benefit of the gearing is less noticeable.

The Ebert transmission, however, is an interesting variation of this system and is shown in Fig. 6.19. A differential gear is employed so that in 'top gear' the transmission is an 'all mechanical' drive. This is achieved by blocking the outlet of a hydraulic pump so that —as in previous versions—it is unable to rotate, and thus brings a pump annulus to rest. As a result the differential gear drives the output shaft in a 1 : 1 ratio.

For torque conversion, the pump is permitted to rotate and its fluid is delivered progressively to three hydraulic motors equal in size to the pump unit. For low torque multiplication motor M_1 alone is employed and its output torque is transmitted to the output shaft via a multiplate clutch. For higher torque multiplication, motor M_2 is then introduced but this drives the output via a 2 : 1 reduction gear so that the motor torque is doubled but can only be employed at lower speed ratios, because motor M_2 rotates at double speed.

Finally, motor M_3 is introduced so that all motors are adding extra torque to the output shaft. This motor drives through a 4 : 1 gear so that as shown in the graph a total output torque multiplication of 8 : 1 is obtained using hydraulic motors of a nominal torque capacity of only 3 : 1. In addition to the reduction in size and weight, this technique employs the motors at more efficient operating regions. As seen in Fig. 6.13 one of the difficulties of vehicle hydraulic transmissions is that the hydraulic motor must be large enough to produce the maximum output torque, but when operating at higher road speeds—top gear—for instance, the motor is operating at very low torque and at a very low efficiency.

Clearly there are many possible systems in which gearing and hydraulic drives may be combined but the overall combination must show some advantage as compared with the 'straight hydrostatic' drive. The advantages which are most desirable are the reduction in size (hence cost) of the hydraulic parts together with an improvement in efficiency. The sharing of power flow between gearing and a hydraulic drive makes it possible to achieve both these advantages provided the extra gearing does not outweigh the reduction in the size of the hydraulic drive.

One example of such a transmission invented by the author is given as Plate 6.7 which shows an aircraft alternator drive. This drive responds quickly to any change in engine speed to maintain an electrical alternator at a required output speed with a very

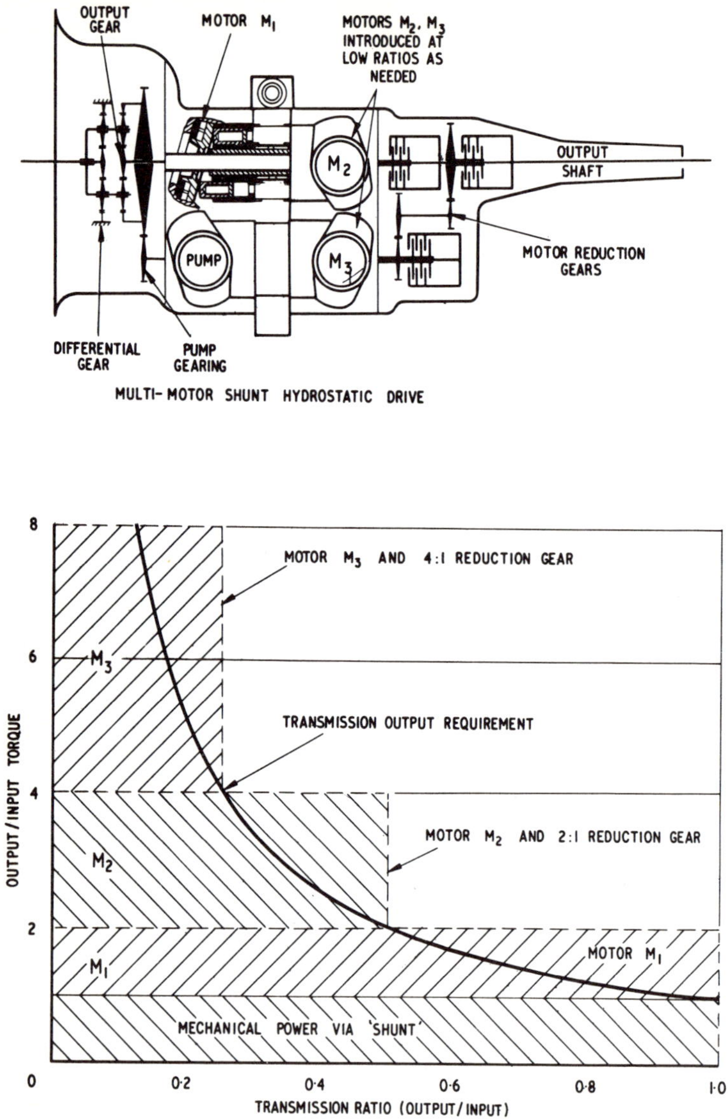

Fig. 6.19 Diagrammatic arrangement of a multi-motor type of shunt hydrostatic drive

small permissible error ($\pm 1\%$). This variable transmission is based on the variable delivery vane pumps shown in Fig. 6.15 and thus operates at a comparatively low pressure. Only 18% of the input power is transmitted by the hydraulic drive and the compact size of this 30 hp drive makes it a strong contender for a small-car transmission and also for other possible applications.

A double differential gear is employed in this device and the hydraulic drive is connected between one member in each differential. By blocking alternately the pump or motor unit, each

Plate 6.7 The Hobson constant speed alternator drive, which is a constant speed alternator drive based on a double differential gear set. The hydraulic drive is employed to hold two control members stationary or to drive them at varying speed. The hydraulic power requirement is low but the speeds of the hydraulic units are extremely high (4–7,000 rev/min). *(By courtesy of H. M. Hobson Ltd.)*

gear set can operate as a simple mechanical drive. The transmission can thus operate as a two-speed gearbox at good efficiency and by employing the hydraulic drive in a normal manner, all the range of ratios between these two fixed speeds—and to some extent outside—can be obtained (see Fig. 6.20).

As seen from the graph the power requirement of the hydraulic

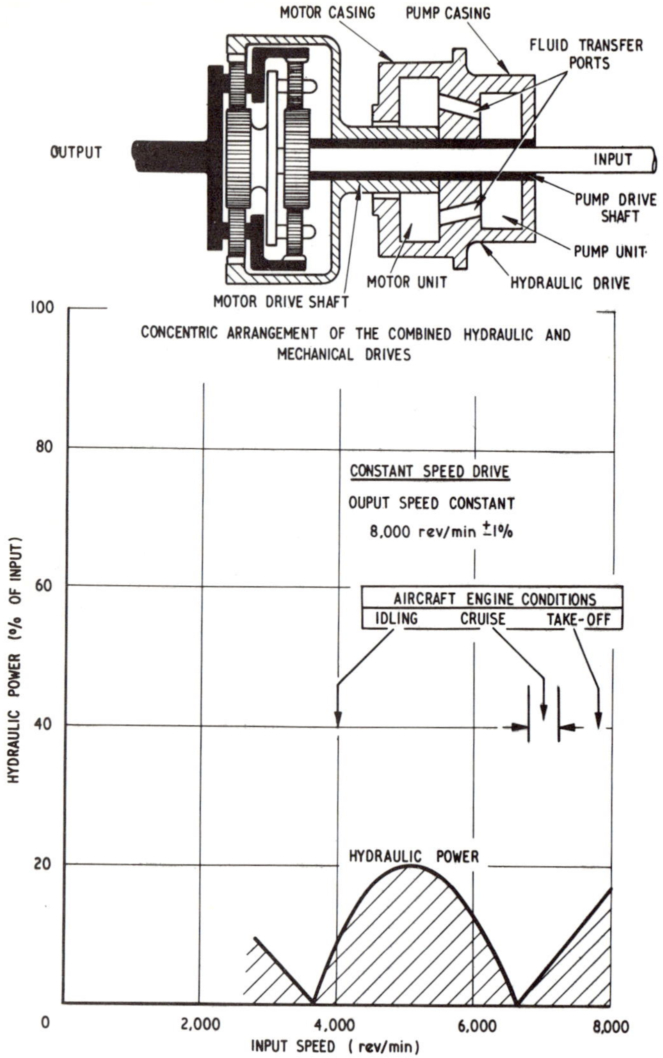

Fig. 6.20 Diagrams illustrating the arrangement of gearing with a hydraulic drive, and the power distribution for a double differential combination

drive is a small proportion of the input power so that there is a good reduction in the size, weight and cost of the hydraulic components, and an improvement in efficiency resulting from the use of the gearing.

6.5 CONCLUSION

It will be clear from the foregoing that some of the devices described in this chapter are in limited use only. A variable ratio drive provides a solution to most of the problems being met with stepped ratio transmissions, but the variable ratio drive is at present in its infancy, has special requirements in achieving a satisfactory automatic control, and may have its own problems. Most engineers believe that road vehicles will ultimately adopt some transmission of this form. The demanding requirements of automobiles in respect of cost and reliability, etc. are such that it is probable that these drives will find success first in other applications before becoming available in a mass production market.

BIBLIOGRAPHY

1. ANON. 'Hayes Transmission', *Automob. Engr.*, **23,** No. 11, 454 (Nov., 1933)
2. FELLOWS, T. G., 'Perbury Transmission for Passenger Cars', Cranfield Symposium (1963). *Advanced School of Automobile Engineering*
3. POON, S. Y. and HAINES, D. J., 'Frictional Behaviour of Lubricated Rolling Contact Elements', *Proc. Inst. Mech. Engrs. (London)*, **181,** 363, Pt. 1, No. 16, (1966/7)
4. MCFARLAND, F. and CHAPMAN, C. S., 'The Buick Flight-Pitch Dynaflow', *S.A.E. Paper* No. 29A, 4–6 March (1958)
5. CHAPMAN, C. S. and GORSKY, R. J., 'The Dual Path Turbine Drive', *S.A.E. Congress,* Paper No. 290B, 9–13 Jan. (1961)
6. MORTENSON, P. C., 'Hydrostatic Transmissions', *S.A.E. National Machinery Meeting,* Paper No. 92U, 14–17 Sept. (1959)
7. THOMAS, J. H., 'Progress in Hydrostatic Transmissions', *Proc. Natn. Conf. Fluid Power,* **19,** 165–172 (1965)
8. MOLLY, H., 'Hydrostatic Vehicle Transmissions. Their Arrangement and Design', *A.T.Z.*, **68,** No. 10, 339–346 (Oct., 1966)
9. SADLER, C. L., 'Differential-type Hydrostatic Transmissions', 1951 *National Conference Industrial Hydraulics,* **5,** 41–50
10. HICKS, R. J., *Variable Ratio Transmission of High Powers using Epicyclic Gears, Allen Engng. Rev.*, No. 48 (Aug., 1962)
11. WESTBURY, R., *et al.*, 'A Double Differential Hydrostatic Drive', *Oil Hydraulic Power Transmission Conference* at Inst. Mech. Engrs. (29 Nov., 1961)

7

vehicle axles

D. Thaw, C.ENG., F.I.MECH.E., M.S.A.E., M.I.MAR.E.,

The function of the rear axle is to support the weight of the rear of the vehicle, to mount the rear, usually driving, road wheels and to transmit power at an appropriate speed from the propeller shaft to the road wheels. There are many configurations suitable for the purpose.

7.1 AXLE HOUSING AND AXLE SHAFT ARRANGEMENTS

Complete rear axle assemblies can be divided into two categories 'Live Axles' and 'Dead Axles'. The latter group are those in which the axle has no part in the transmission of power and merely supports the weight of the rear of the vehicle. They are largely obsolete. The live axles can be further classified as semi-floating, three-quarter floating and fully floating. Fig. 7.1 shows examples of these categories.

It will be seen from the illustration of the fully floating axle that the axle shafts are subjected only to torsional stress. The axle housing extends through the wheel and this is carried on bearings mounted on the axle housing. The entire reaction from the wheel is therefore transferred to the axle housing rather than to the axle shaft in the form of a bending moment. Advantages of this form of construction are a reduction in axle shaft principal stress, the ability to remove a broken axle shaft without removal of the wheel, and the fact that the vehicle remains towable if an axle shaft breaks. The fully floating axle is virtually standard for truck applications.

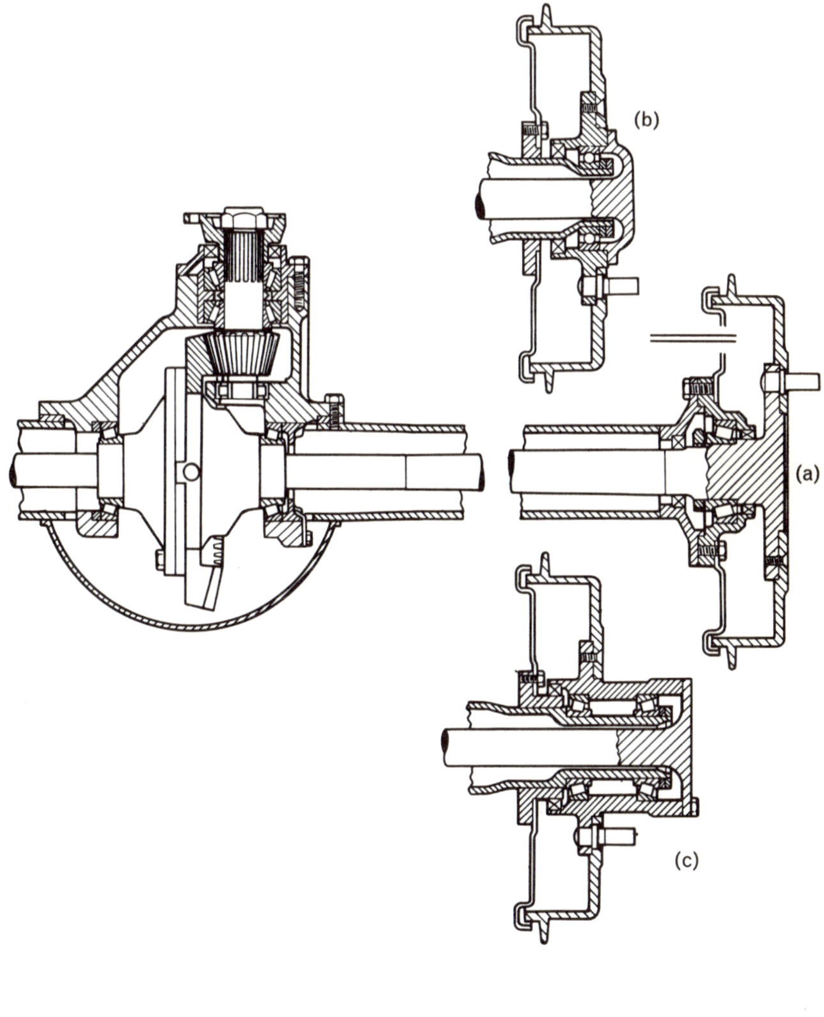

Fig. 7.1 Basic classification of axles: (a) semi-floating; (b) three-quarter floating; (c) fully-floating

The semi-floating axle is the simplest in construction. Here the axle shaft is subjected to torsion from the drive and to bending moment from both the weight of the vehicle and from all other wheel reactions. It is the most popular construction for passenger cars.

The three-quarter floating axle is midway between the previous two. If the outer bearing is located on the vertical centre line of the road wheel, the axle shafts are relieved of the bending moment due to the weight of the vehicle. A bending moment is still set up in these components due to lateral and fore and aft reaction from the wheel. This construction is sometimes employed in light van and some passenger vehicles.

7.1.1 AXLE HOUSING TYPES

It will be seen that a centre unit comprising the bevel reduction and differential assembly is bolted to the axle housing proper. The axle shafts driven by splines in the differential side gears run within the tubes of the axle housing, their outer flanges connected to the brake drums and the road wheels. The outer ends of the axle housing tubes have flanges on which are carried brake back plates. Between these flanges and the centre portion of the axle housing, spring seats, hand brake and shock absorber anchorages are welded.

The axle housing may be built up in different ways. For passenger vehicles a cast centre section is favoured which encloses the gearing and has separate tubes assembled into it to extend the housing to the desired track width. For heavier duty applications the housing may be formed completely from one tube. The centre portion of the banjo is formed to a larger diameter by a series of swaging operations and the side tubes reduced in size by hot pressing. This is an excellent method since the final housing is of integral construction and the necessary pressing operations thin down the metal in the enlarged centre section and increase the thickness of the side tubes where extra strength is needed to carry the weight of the vehicle. For large axle housings, beyond the capacity of the above process, two identical pressings are used which are welded together along the horizontal centre line of the housing.

7.2 THE PASSENGER CAR AXLE

The most common driving head layout for passenger cars is shown in Fig. 7.2, where the input flange of the driving head is coupled

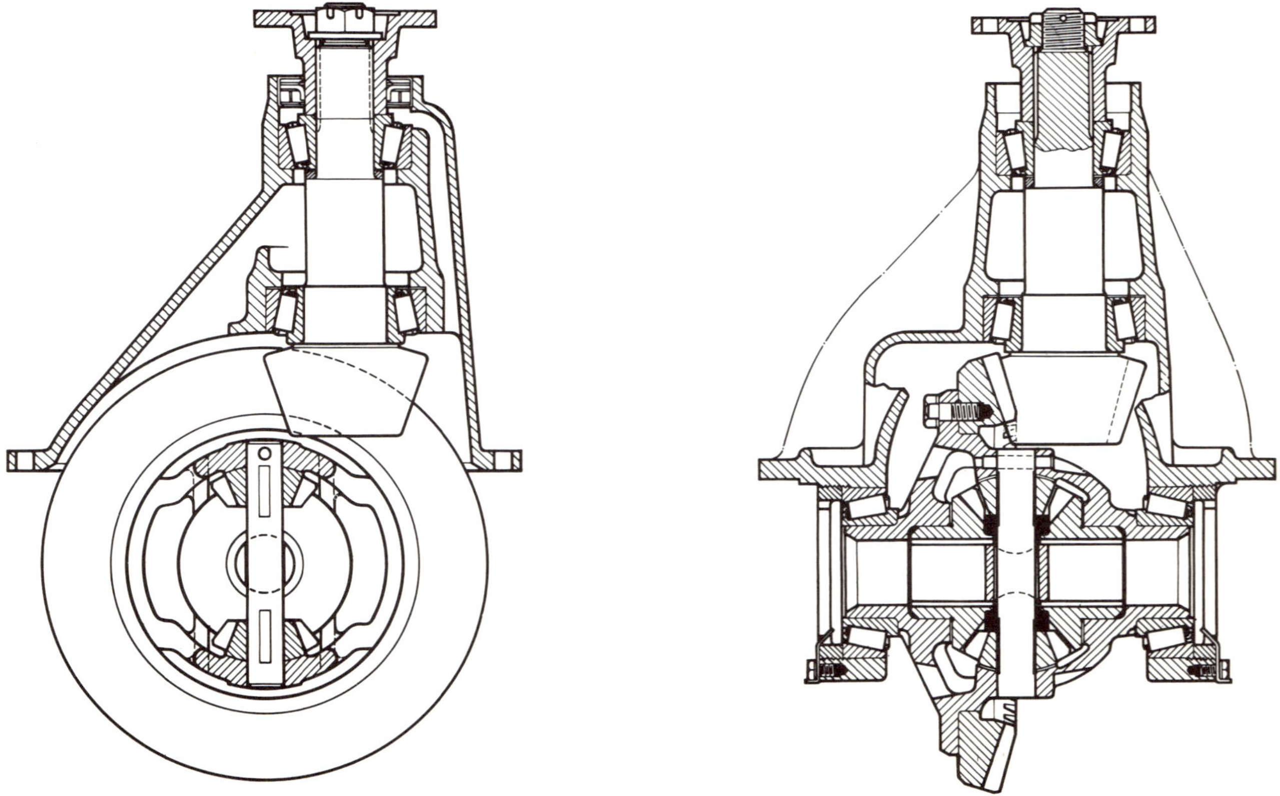

Fig. 7.2 A typical passenger car axle

directly to the propeller shaft and a right angle drive is provided to transmit speed and torque to the road wheels. The requirement for a right angle drive capable of transmitting high torques but of only modest dimensions is met by bevel gearing, either spiral or hypoid. The special advantages of the hypoid gear have led to its almost universal adoption in passenger car axles.

Plate 7.1 shows typical crown wheel and pinion pairs of the spiral bevel and hypoid configuration. The main difference between them is that whereas the spiral bevel pinion is mounted on the horizontal centre line of the crown wheel, the hypoid pinion has an offset, usually below the centre line of the crown wheel. Two benefits follow from this offset. From the vehicle point of view the lowering of the drive line removes or reduces the incursion of the transmission tunnel into the floor of the vehicle. From the gearing point of view the geometry of the offset allows a larger diameter and therefore stronger pinion than in a spiral bevel pair of the same crown wheel diameter and ratio. This advantage can be used to provide stronger gearing with the same crown wheel size as for a spiral bevel pair, or conversely gearing with equal strength having a smaller crown wheel size. The offset also leads to higher tooth spiral angles, which by giving increased helical overlap promote quietness and smoothness in operation. These advantages are offset by higher tooth loadings and increased sliding velocities along the teeth. With the use of modern surface treatments and Extreme Pressure lubricant additives, these do not present a serious problem.

The design, manufacture and development of the spiral bevel or hypoid gearing remains as much an art as a science. A certain amount of predictable change in tooth helix and form is encountered during the heat treatment process. It is necessary, therefore, to adjust the cutting in the soft stage to ensure that the required form is obtained after subsequent heat treatment. Gears of this type are extremely sensitive to correct mounting and their performance is greatly influenced by the deflection of the casings in which they are mounted and by the relaxation of the bearings locating them. If the gears were produced to give a full length and full profile contact when tested in a rigid and accurate test machine, these conditions would not be realised in subsequent running in the production axle. The effect of casing deflection and of bearing relaxation is to concentrate the load at one or other end of the gear tooth, leading to its early destruction. It is therefore necessary to ensure the greatest rigidity of the casing, and of the bearing arrangement. Even so, the gears have to be produced with a localised tooth marking which, under the influence of normal

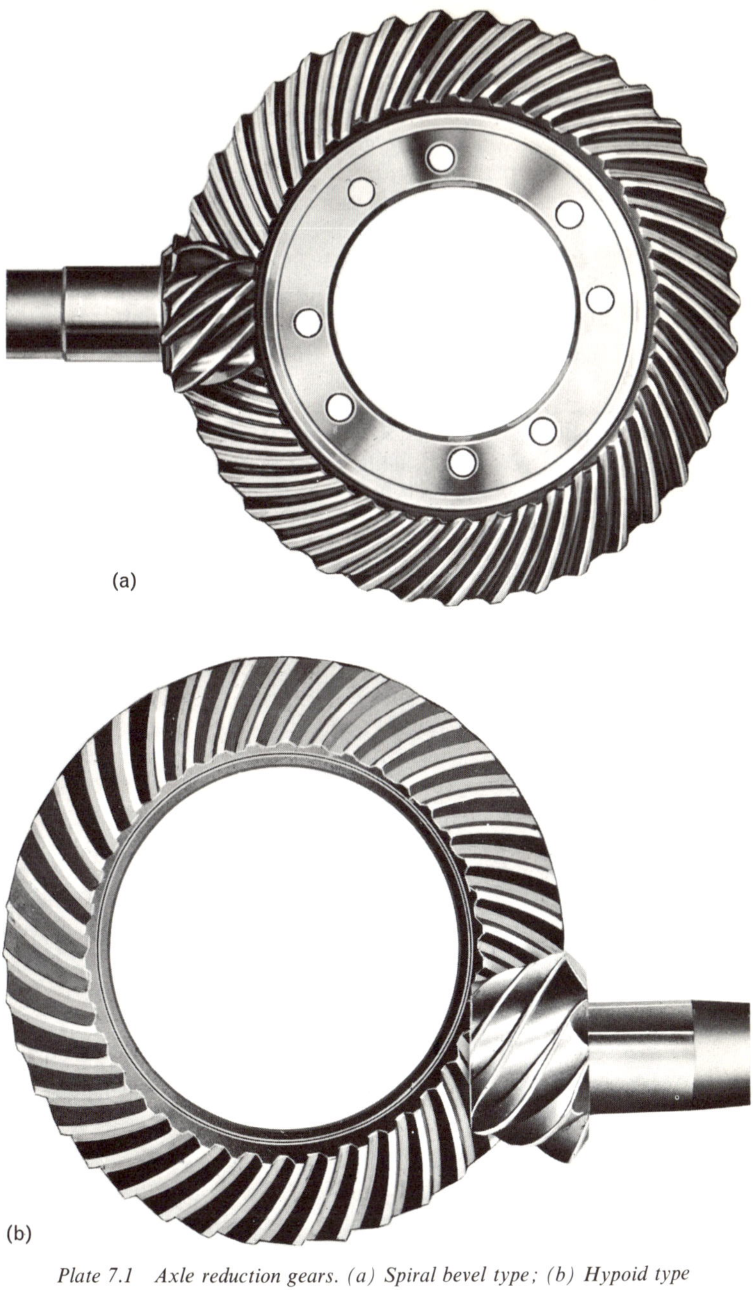

Plate 7.1 Axle reduction gears. (a) Spiral bevel type; (b) Hypoid type

deflection and bearing relaxation, will spread to give the desired full length and full profile marking.

It is desirable in every case, and essential where high production is anticipated, to carry out a full scale deflection test of the entire driving head. In such a test the gears are motored slowly and the spread of the tooth marking under successive increments of torque is observed. In addition, indicator clocks are mounted to record the movement of the gears and to establish the rigidity of the structure and of the bearing arrangement. As shown in Fig. 7.2 opposed taper roller bearings are commonly used to ensure the necessary rigidity of mounting. Since some settling down is usual during the early mileage of the axle, it is usual to introduce a pre-load into this opposed bearing arrangement. Such pre-load is usually specified as a particular friction torque in lb in at the input shaft of the axle assembly and is obtained by carefully adjusting by shims the position of one bearing race relative to the other. The frictional resistance rapidly reduces in early running but the gearing remains in positive location.

7.3 THE DIFFERENTIAL

Common to all modern rear axles is the differential and a short note on its history and uses may be appropriate. The differential is thought to have been invented in 1827 by Onesiphore Pecqueur. Prior to this, rigid axles had been used in which both drive wheels had been constrained to run at the same speed. When cornering, each drive wheel must cover a different path in the same interval of time and the use of the rigid axle led to skidding and slippage of one drive wheel relative to the other. A mechanism was required which would transmit torque to each road wheel whilst allowing each to run at a speed appropriate to the path that it had to travel. The invention of the differential solved the problem.

A typical bevel differential is shown in Fig. 7.3. The two large bevels *A* and *B* are termed side gears and their splined bores form the drive connection with similar splines on the inner end of the axle shafts *C* and *D*. Meshing with the side gears are the differential pinions *E* and *F*.

Depending upon the total torque to be transmitted there may be any convenient number of differential pinions. In practice two are usually found sufficient for passenger cars and four for commercial vehicles. Where two are used they are mounted and free to rotate on the cross pin *G* which is firmly located at each end in the differential casing *H*. Where four such pinions are necessary two intersecting cross pins are used and they commonly terminate in an annular ring in the centre to form a 'spider'. If we examine Fig. 7.3, it will be observed that the bevel crown wheel *J* is bolted to a flange on the differential casing and drives the entire unit. When the vehicle follows a straight path both road wheels and their associated axle shafts and the side gears rotate at the same speed. The

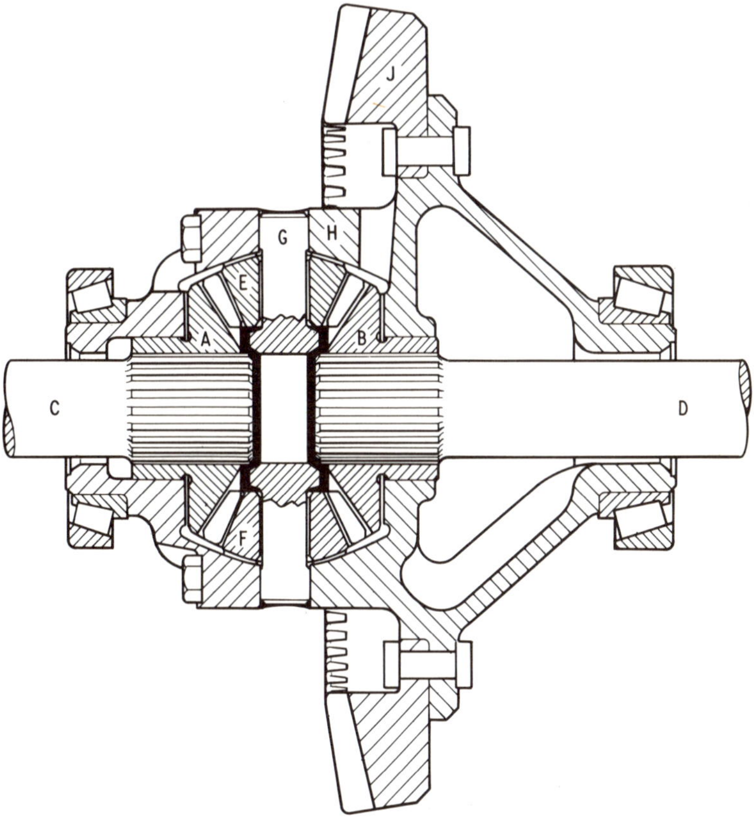

Fig. 7.3 Bevel differential gear

torque is transmitted through the crown wheel to the differential casing and by way of the cross pins to the differential pinions. These impress equal torques on the two side gears, which torques are transmitted to the two rear driving road wheels via the axle shafts. In straight line running there is no relative motion between the differential pinions and the cross pin or the side gears. The entire differential rotates en bloc and fulfils the function of the rigid toothed coupling between the two axle shafts. When the vehicle negotiates a curve the outer wheel follows a longer path than that of the inner wheel and must therefore rotate at higher speed. In this condition the differential, as it must, still impresses equal torque on each axle shaft but the difference in speed of the two axle shafts results in rotation of the differential pinions on the cross pin. The angular speed of the differential casing is always equal to the algebraic mean of the speeds of the two side gears.

7.3.1 THE LIMITED SLIP DIFFERENTIAL

One of the virtues of the rigid axle, and one of the shortcomings of the differential, is currently becoming more obvious. If one drive wheel encounters a patch of ice or mud and has insufficient adhesion it will slip. Since the differential at all times impresses equal torques on the two road wheels it can impress no greater torque on the wheel with adhesion than can be transmitted by the wheel which has lost adhesion. The wheel on firm ground will remain stationary, the wheel without adhesion spinning at twice its normal speed. The increasing power to weight ratio of the modern passenger car results in lower adhesion factors at the drive wheels, and this shortcoming of the differential is becoming a nuisance under conditions of rapid acceleration and of weight transference during cornering.

Many devices have been designed to prevent loss of traction when one driving wheel loses adhesion. Several off-highway vehicles incorporate a differential lock in which a clutch locks two elements of the differential which then becomes a rigid coupling. When used with care this device is most effective, although if the differential remains locked during cornering on normal roads tyre scrub and high axle shaft torques are generated. In some cases tapered dog clutches disengage the lock at a predetermined torque.

For passenger cars the differential lock is too severe a solution and the principle of the majority of devices used is the provision of a frictional drag torque for the side gear associated with the slipping road wheel. By this means some torque at least can be transmitted to the road wheel having adhesion.

Such a device is shown in Fig. 7.4 which illustrates a differential equipped with the Eaton 'Trac-Aide' limited slip feature. Interposed between the side gears and the differential casing are a number of pairs of specially textured clutch discs, one of each pair held stationary at its periphery by the differential casing, the other driven by the splined hub of the side gear. In straight line running there is no relative rotation between these pairs of clutch discs. However in all conditions of a difference in speed between the two side gears relative motion is set up between each pair of discs. The inherent gear thrust from the bevel side gears provides an axial loading of the clutch pack which then generates a frictional drag

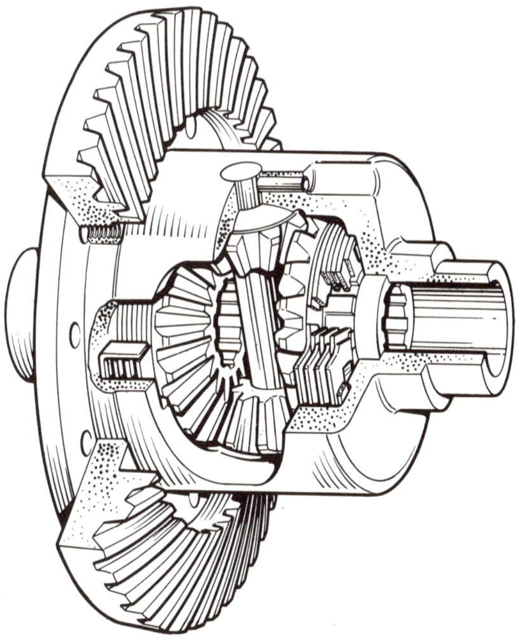

Fig. 7.4 Eaton 'Trac-aide' limited slip differential gear

torque on the side gears. The axial load is frequently augmented by helper springs interposed between the two side gears. In the most extreme condition, a condition resulting in complete loss of adhesion on one drive wheel, a certain friction torque is impressed on the appropriate side gear by the axially loaded clutch pack thus permitting a torque of equal magnitude to be transmitted to the road wheel on firm ground. This beneficial effect is noticeable in all milder conditions of unequal road wheel adhesion.

Fig. 7.5 shows another example of the limited slip differential, the 'Powr-Lok', which again is based upon the friction torque generated by the axial loading of a clutch plate pack. In this case the axial load is provided from three separate sources. As before the differential bevel gearing provides an axial thrust which is supplemented by springs, this time in the form of Belleville

Washers. In addition the differential cross pins are mounted in 'V'-shaped cams in the differential housing. Under drive torque the cross pins tend to climb these inclined ramps thus providing a further axial force for the loading of the clutch plate packs (see also Plate 7.2).

7.4 TRUCK AXLES

The truck transmission demands a careful study of overall ratio and the number of ratios available within that overall range. It is not unusual for passenger cars to have a power to weight ratio of

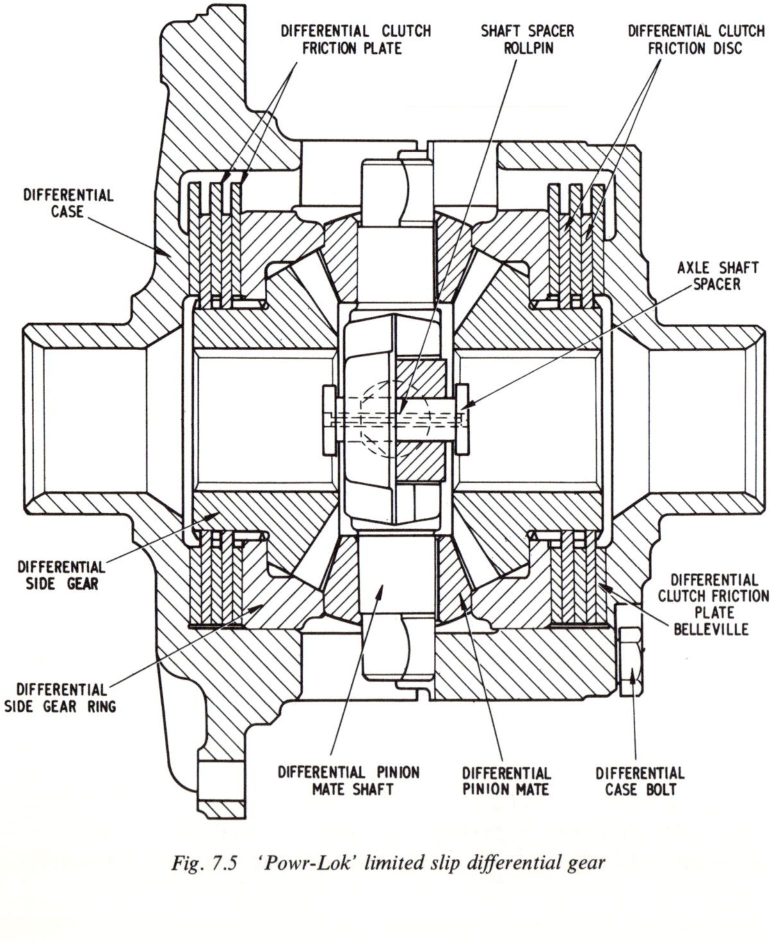

Fig. 7.5 'Powr-Lok' limited slip differential gear

80 bhp/ton. In these circumstances unless the ultimate in performance is required three ratios are sufficient for normal driving and the vehicle will remain flexible over a wide speed range in each gear. The truck, however, has a much lower power to weight ratio, of the order of 6–7 bhp/ton. In order to obtain an acceptable

Plate 7.2 'Powr-Lok' limited slip differential gear. (By courtesy of Salisbury Transmissions Ltd.)

performance the relative lack of power must be compensated by the provision of a greater number of gear ratios so that in any condition of operation a ratio is available which allows the engine to work in a narrow speed band which gives high torque and efficiency.

The vehicle manufacturer has the problem of deciding what overall reduction ratio is necessary between the engine and the driving wheels, and how and where this overall reduction should be achieved. At the maximum governed speed of the engine the gearing should give the vehicle a top speed of 55–60 m.p.h. At the other end of the range the gearing should give the vehicle a capability of climbing at least a 20% (1 in 5) grade. To achieve this kind of performance the British medium truck requires an overall ratio of about 55/1. With the advent of the high speed diesel engine this overall ratio is likely to increase and may well approach 80/1 in the future. Since the maximum economical first gear ratio in the gearbox is about 8·5/1 and the maximum single step through the axle bevel gearing is about 7/1, it is clear that a single-speed axle is barely adequate for present use and will certainly be inadequate for future vehicles.

These considerations affect the design of the rear axle, and various combinations of gearbox and axle are emerging.

1. 5-speed gearbox and single-speed axle.
2. 8- or 10-speed gearbox and single-speed axle.
3. 5-speed gearbox and 2-speed axle.
4. 5-speed gearbox and 3-speed tandem axle.

7.4.1 THE RATIO CHART

It is convenient to show the effect of an increased number of available ratios by drawing Ratio Charts and Figs. 7.6 and 7.7 show respectively the use of a 5-speed gearbox and single-speed axle, and a 5-speed gearbox and 3-speed tandem axle. These represent almost the extremes of acceptable ratio steps. It will be clear that, at least from the vehicle operating point of view, the 15-speed combination is much more efficient. Within these two extremes there is a progressive improvement with the increasing number of ratios available.

In operation the truck is accelerated from rest to an engine speed of say 2,000 rev/min. The first sloping line *OA* in Fig. 7.6 shows this acceleration and indicates a road speed of 6 m.p.h. at the end of the acceleration period. For acceleration beyond this speed the driver shifts into second gear which pulls down the engine speed. This is indicated in Fig. 7.6 by the vertical line *AB*. The engine is again accelerated to 2,000 rev/min indicated by sloping line *OC* and this procedure is repeated for third, fourth and fifth gears producing sloping lines *OD*, *OE*, *OF*. The horizontal line from 1,500 rev/min represents the maximum torque speed of the engine.

Fig. 7.7 is constructed upon exactly the same principles and indicates the state of affairs when fifteen ratios are available from the axle/gearbox combination.

In all operation other than high speed or motorway running

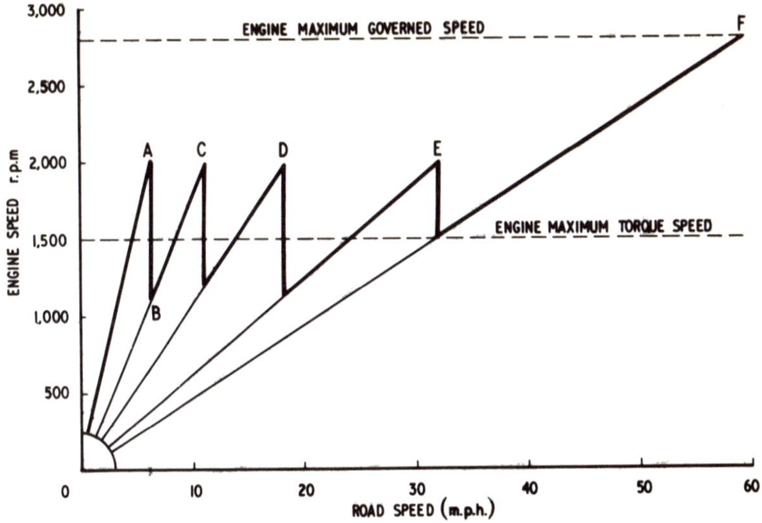

Fig. 7.6 Engine speed and ratio chart for a vehicle employing a 5-speed gearbox and single-speed axle

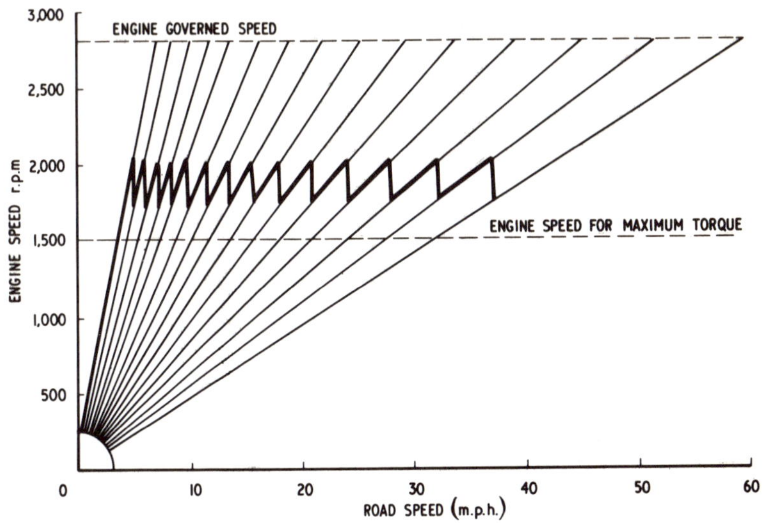

Fig. 7.7 Engine speed and ratio chart for a vehicle employing a 5-speed gear box and 3-speed tandem axle

there is a fairly narrow band of engine speed where the balance of fuel economy and engine torque available results in the most economical use of the engine. On a typical engine this speed band is from 1,300 to 1,600 rev/min. The vertical lines in Figs. 7.6 and 7.7 show engine speed variations in the course of normal gear changing and it is clear from these diagrams that more economical operation is possible with the higher number of ratios available. An incidental, but important benefit from the provision of more ratio steps is that the gap between each step becomes smaller and that gear shifting becomes easier on this account.

7.4.2 THE SINGLE SPEED AXLE

Plate 7.3 shows a typical single speed truck axle. Since the lower transmission line is not required and silence is less important than

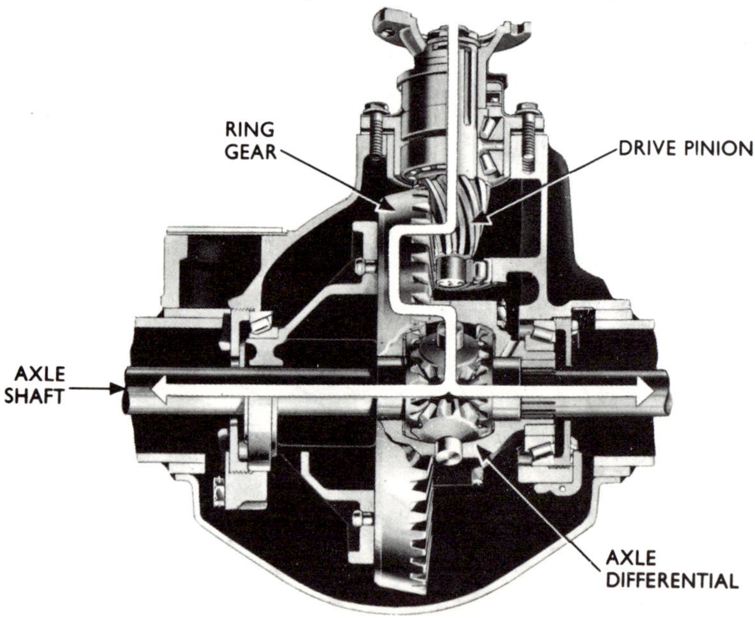

Plate 7.3 A typical single speed truck axle

in the passenger vehicle, truck axles are usually based on the use of spiral bevel gearing. Operating torques are high and truck usage subjects the unit to high impact loading. In these circumstances particular care is necessary to prevent large deflections of the casing and bearings from affecting the contact conditions of the spiral bevel gears. It will be noticed that the bevel pinion of the passenger car axle in Fig. 7.2 is overhung from its bearings. The truck axle illustration shows the introduction of a 'straddle' bearing to give maximum support to the bevel pinion. The nose-piece casting is also much heavier and well ribbed to contain the loads imposed upon it with the minimum of deflection.

Lubrication of the unit is initiated by the crownwheel dipping into the oil reservoir in the bowl of the banjo. The oil is carried

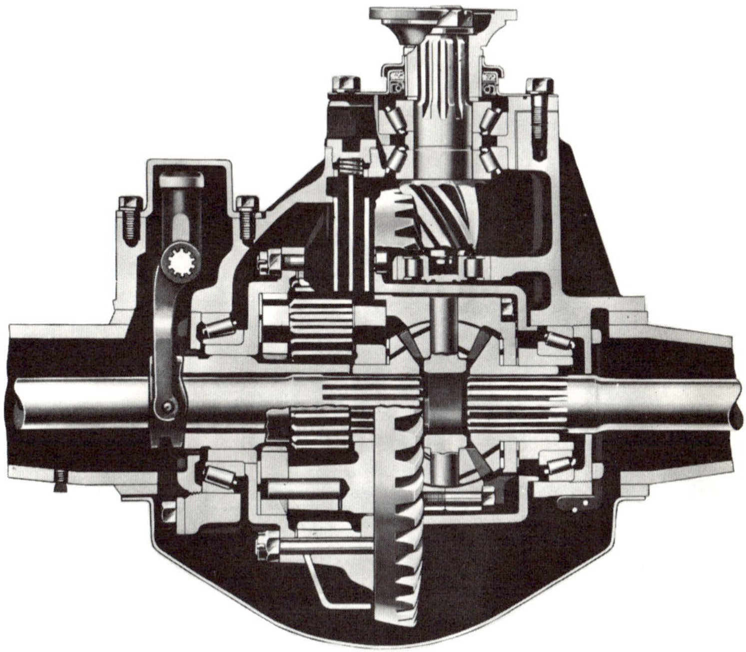

Plate 7.4 The Eaton 2-speed axle. (By courtesy of Eaton Axles Ltd.)

round to the meshing point of the spiral bevels and is splashed into the differential and the crown wheel bearings. Special cored passages guide the oil flung from the periphery of the crown wheel into the pinion bearing assembly to ensure adequate lubrication.

7.4.3 THE TWO SPEED AXLE

The combination of 5-speed gearbox and 2-speed axle is very popular. Plate 7.4 shows the Eaton 2-speed axle which in principle is a single speed axle combined with the epicyclic reduction gear which can be controlled to act as a direct drive or to give a reduction of 1·38 to 1. Thus in high range the axle has a ratio equal to the ratio of the spiral bevels and in low range this ratio is multiplied by 1·38. A typical range of ratios which can be supplied within the same casing is given in Table 7.1.

Plate 7.5a shows the axle in high ratio; the sun pinion clutch is in mesh with the planetary pinions and is engaged with the high-

Table 7.1

High Range	6·5	6·14	5·57	5·14	4·87	3·83
Low Range	9·04	8·54	7·75	7·15	6·77	5·32

Plate 7.5a (a) 2-speed axle in high ratio. (By courtesy of Eaton Axles Ltd.)

speed clutch plate, preventing the planetary pinions from rotating. The unit therefore functions as a conventional single reduction axle, the epicyclic being inoperative and acting merely as a coupling.

In Plate 7.5b the sun pinion clutch gear has been shifted into mesh with the low-speed clutch plate and the planetary pinions;

Plate 7.5b (b) 2-speed axle in low ratio. (By courtesy of Eaton Axles Ltd.)

these now rotate about the clutch gear which is held stationary. A secondary reduction is therefore introduced between the crown wheel and the differential.

The shift control is from a button on the gear lever and works through an electrical or air pressure control mounted on the axle. An adaptor changes gear in the speedometer drive at the same time as the axle ratio changes so that the speedometer reading is corrected to the changed axle ratio and true road speed is always indicated.

Great care is paid to the lubrication of the unit. Oil is picked up from the banjo reservoir by a collector drum behind the crown wheel, the lubricant is then caught by a scoop and divided into two ducts, one feeding the pinion bearings and the other the planetary gears and differential bearings. Oil is supplied to these vital points within half a turn of the road wheels and thereafter the flow is continuous while the vehicle is in motion.

Fig. 7.8 shows the Rockwell Standard 2-speed axle. This has a first reduction pair of spiral bevel gears and two second reduction pairs of single helical gears. The pinions of both sets of helical gears are loose on the intermediate shaft but can be drivably connected with this shaft by means of an internally splined driving clutch, the teeth of which engage with similar teeth on the barrel portion of the intermediate shaft and with teeth formed integrally at the inner end of the pinions. Movement of the internally splined driving clutch axially in either direction gives a choice of two speeds at the axle output. Where a single-speed low ratio axle is required both of the axles described above can be made in such a way that only the low ratio is operated. This leads to considerable simplification as no shift mechanism or controls are necessary.

7.4.4 TANDEM AXLE DRIVES

In vehicle design an effort is made to attain correct load distribution or placement of gross weight in relation to tyres and axles. A desirable load distribution for a four-wheel truck is $66\frac{2}{3}\%$ on the rear axle and $33\frac{1}{3}\%$ on the front axle. Thus a vehicle with 24,000 lb gross laden weight should ideally carry 16,000 lb on the rear and 8,000 lb on the front. Such a distribution is rarely, if ever, attained in practice.

The situation is more favourable on the six-wheeled vehicle. Using a 40,000 lb six-wheeled truck as an example, 16,000 lb can be placed on each of the rear axles with only 8,000 lb on the front axle. This is good weight distribution and is commonly achieved in practice. This correct weight distribution leads to increased tyre life and the two extra brakes incorporated in the third axle lead to increased safety margins and reduce the likelihood of skidding.

Wider adoption of the three-axle vehicle has led to the Eaton Tandem Drive Unit, an external view of which is shown in Plate 7.6. These tandem axles can be supplied as single reduction or as 2-speed units. The 2-speed unit is chosen for this description. In the tandem unit both axles are driven and each must have its own driving head. The driving head of the rear of the two axles is

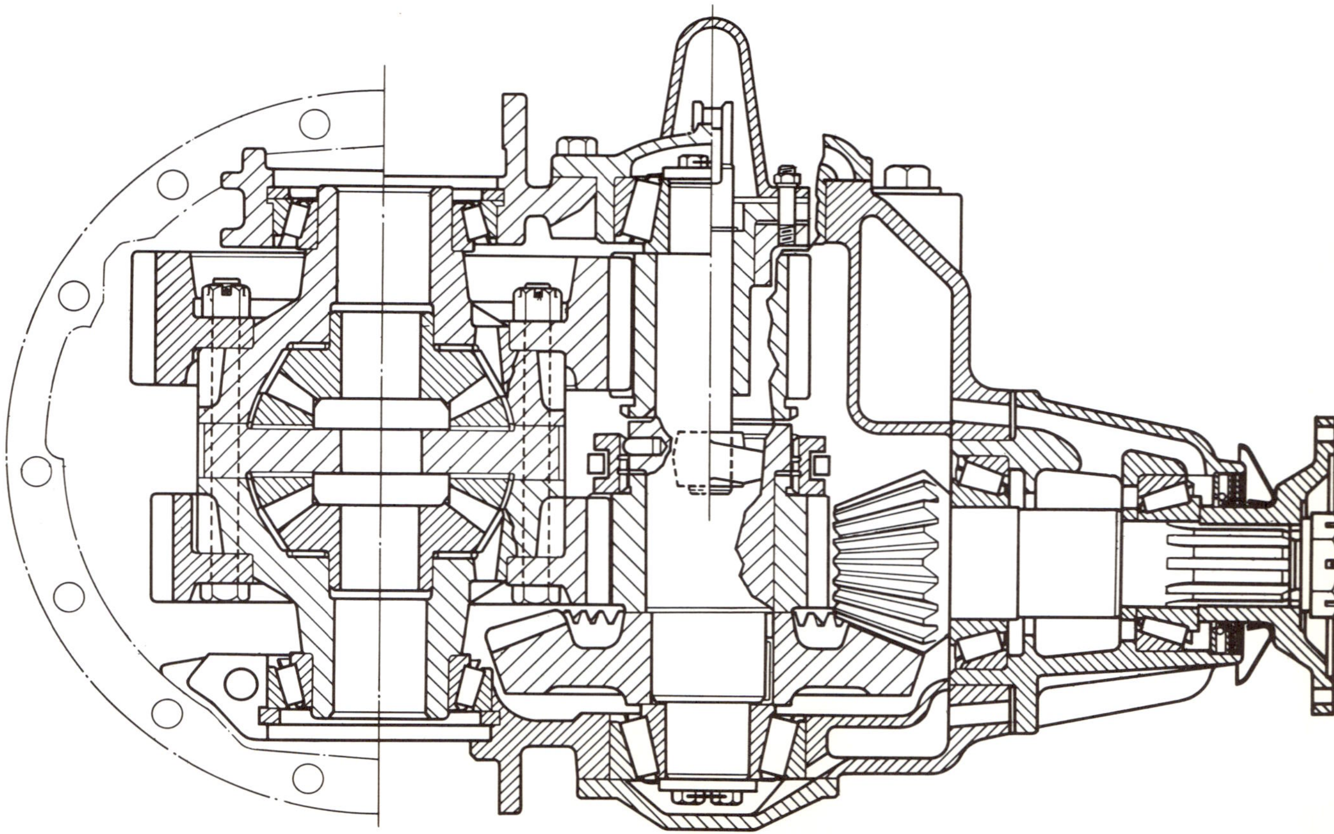

Fig. 7.8 Rockwell Standard 2-speed axle

Plate 7.6 Eaton Tandem Drive unit. (By courtesy of Eaton Axles Ltd.)

exactly the same as the standard two-speed assembly described previously. Power has to be passed through the front driving head to the rear and hence a change of design is necessary in this unit. Plate 7.7 shows both the forward and rear axle driving head. In the forward driving head the input drive torque is transmitted to a differential, the 'third differential'. The function of this unit is to divide power equally between the front and rear axles of the tandem in all operating conditions and prevent a fight between the two axles due to differences in tyre condition and terrain. With three differentials in the tandem assembly all four rear wheels are allowed to travel at different speeds for cornering and yet have equal drive torques impressed on them. When operating in off-highway conditions the third differential can be locked out by a dog clutch to produce a positive drive to each axle for maximum traction. Operation of the range shift is again from a button control in the cab and synchronous shifting of the two axles is automatically obtained. Tracing the power flow from the third differential shows that this unit has one input, from the propeller shaft, and two outputs. Output from one side of the differential drives the bevel pinion of the forward axle via single helical gears. Output from the other side of the differential provides a drive to the input of the rearward axle.

7.4.5 THE THREE SPEED TANDEM

Although the third differential is an essential component in the tandem axle layout its potential is not being fully utilised. The ability of a differential mechanism to divide torque equally between two units regardless of the speed difference between them has been utilised in tandem axles, to coax three ratios from what is essentially a two speed ratio axle arrangement. Thus in addition to the two ratios available with both axles in high or in low range it is also possible to obtain an intermediate ratio with one axle in high range and the other in low range. In this application the third differential distributes not only equal torques but appropriate input speeds to the forward and rearward axles and its duties are much more onerous than when used with the 2-speed tandem axle. Special care is necessary in the mounting of the differential gears and generous lubrication is essential.

7.4.6 HUB REDUCTIONS

As indicated in Section 7.4 the overall speed reduction required from engine to road wheels is steadily increasing. One way to

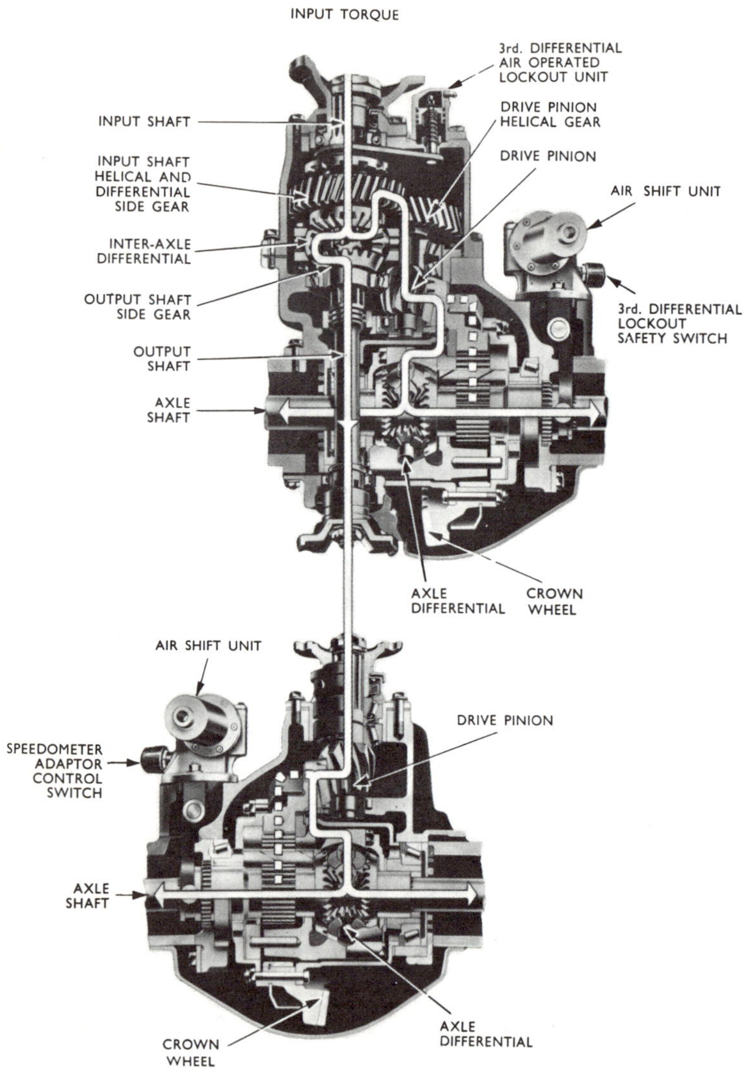

Plate 7.7 Sectional view of the Eaton Tandem Drive Unit. (By courtesy of Eaton Axles Ltd.)

provide a fixed reduction ratio is by the use of hub reduction units. There is much to be said for locating a final reduction at the wheels. In the normal arrangement of gearbox and rear axle the propeller shaft must be proportioned to transmit the heavy torque associated with the required low output speed from the gearbox. Similarly the rear axle must increase in size because of the higher overall reduction ratio of the transmission. If, for example, hub reductions having a reduction of 2 to 1 are used then this halves the torque to be transmitted by the axle shafts, differential, bevel gears, propeller shaft and gearbox.

Fig. 7.9 shows a typical hub reduction unit which is of fixed ratio epicyclic layout. The input is driven by the axle shaft and the output

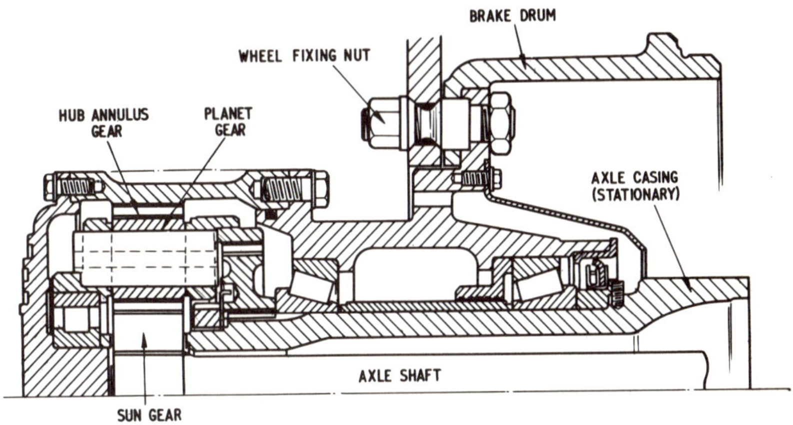

Fig. 7.9 Hub reduction type of axle

drives the road wheel. Compact design is dictated by the necessity to keep the outside diameter of the unit small enough to allow assembly of the wheel disc.

BIBLIOGRAPHY

1. *Bevel Gear Manufacture,* Gleason Works, Rochester, N.Y., U.S.A.
2. *How to test Bevel Gear,* Gleason Works, Rochester, N.Y., U.S.A.
3. WELLS COLEMAN, 'Design and manufacture of Spiral Bevel and Hypoid Gears for heavy duty Axle Drives', *S.A.E. Trans.* (Jan., 1962)
4. HELDT, P. M., *The Automotive Chassis,* Iliffe Books Ltd. (1952)
5. NEWTON AND STEED, *Motor Vehicle, Part* 3, Iliffe Books Ltd. (1966)

6. GORDON, K. W., 'Design evaluation and selection, of Heavy Duty Rear Axles.' *S.A.E. Trans.* (Nov., 1954)
7. *Design data for Automotive drive axles,* Gleason Works, Rochester, N.Y., U.S.A.
8. BURTON, HINDLIN and MCDOWELL, 'Gearing the Vehicle to the Road', *S.A.E. Trans.* (Aug., 1961)
9. DEAN-AVERNS, R., *Automobile Chassis Design*, Iliffe Books Ltd. (1952)

Index